Bio- and Nano-sensing Technologies for Food Processing and Packaging

Food Chemistry, Function and Analysis

Series editors:
Gary Williamson, *Monash University, Australia*
Alejandro G. Marangoni, *University of Guelph, Canada*
Graham A. Bonwick, *AgriFoodX Limited, UK*
Catherine S. Birch, *AgriFoodX Limited, UK*

Titles in the series:
1: Food Biosensors
2: Sensing Techniques for Food Safety and Quality Control
3: Edible Oil Structuring: Concepts, Methods and Applications
4: Food Irradiation Technologies: Concepts, Applications and Outcomes
5: Non-extractable Polyphenols and Carotenoids: Importance in Human Nutrition and Health
6: Cereal Grain-based Functional Foods: Carbohydrate and Phytochemical Components
7: Steviol Glycosides: Cultivation, Processing, Analysis and Applications in Food
8: Legumes: Nutritional Quality, Processing and Potential Health Benefits
9: Tomato Chemistry, Industrial Processing and Product Development
10: Food Contact Materials Analysis: Mass Spectrometry Techniques
11: Vitamin E: Chemistry and Nutritional Benefits
12: Anthocyanins from Natural Sources: Exploiting Targeted Delivery for Improved Health
13: Carotenoid Esters in Foods: Physical, Chemical and Biological Properties
14: Eggs as Functional Foods and Nutraceuticals for Human Health
15: Rapid Antibody-based Technologies in Food Analysis
16: DNA Techniques to Verify Food Authenticity: Applications in Food Fraud
17: Advanced Gas Chromatography in Food Analysis
18: Handbook of Food Structure Development
19: Mitigating Contamination from Food Processing
20: Biogenic Amines in Food: Analysis, Occurrence and Toxicity
21: Nutrition and Cancer Prevention: From Molecular Mechanisms to Dietary Recommendations
22: Health Claims and Food Labelling
23: Nutraceuticals and Human Health: The Food-to-supplement Paradigm
24: Nutritional Signalling Pathway Activities in Obesity and Diabetes
25: The Chemistry and Bioactive Components of Turmeric
26: Foodomics
27: Food Proteins and Peptides: Emerging Biofunctions, Food and Biomaterial Applications
28: Handbook of Antioxidant Methodology: Approaches to Activity Determination
29: Fats and Associated Compounds: Consumption and Human Health

30: Oral Processing and Consumer Perception: Biophysics, Food Microstructures and Health
31: Development of Trans-free Lipid Systems and their Use in Food Products
32: Advanced Spectroscopic Techniques for Food Quality
33: Berries and Berry Bioactive Compounds in Promoting Health
34: Metabolism of Nutrients by Gut Microbiota
35: Bio- and Nano-sensing Technologies for Food Processing and Packaging

How to obtain future titles on publication:
A standing order plan is available for this series. A standing order will bring delivery of each new volume immediately on publication.

For further information please contact:
Book Sales Department, Royal Society of Chemistry, Thomas Graham House, Science Park, Milton Road, Cambridge, CB4 0WF, UK
Telephone: +44 (0)1223 420066, Fax: +44 (0)1223 420247,
Email: booksales@rsc.org
Visit our website at www.rsc.org/books

Bio- and Nano-sensing Technologies for Food Processing and Packaging

Edited by

Ashutosh Kumar Shukla
Ewing Christian College, India
Email: drakshukla@gmail.com

ROYAL SOCIETY OF CHEMISTRY

Food Chemistry, Function and Analysis No. 35

Print ISBN: 978-1-83916-432-3
PDF ISBN: 978-1-83916-796-6
EPUB ISBN: 978-1-83916-797-3
Print ISSN: 2398-0656
Electronic ISSN: 2398-0664

A catalogue record for this book is available from the British Library

The Royal Society of Chemistry is a charity, registered in England and Wales, Number 207890, and a company incorporated in England by Royal Charter (Registered No. RC000524), registered office: Burlington House, Piccadilly, London W1J 0BA, UK, Telephone: +44 (0) 20 7437 8656.

Visit our website at www.rsc.org/books

Printed in the United Kingdom by CPI Group (UK) Ltd, Croydon, CR0 4YY, UK

Preface

Safe processing and packaging of food items is undoubtedly in everyone's interest. Sensors have very important roles in processing and packaging. Accordingly, sensing technologies have gained societal importance. Bio- and nano-sensing technologies for processing and packaging applications have attracted the attention of the scientific community in view of increasing environmental concerns. This edited volume presents a collection of 10 chapters discussing the current trends of bio- and nano-sensing technologies for processing and packaging of food items. The first chapter, Sensors for the Food Industry: An Introduction by D. Cozzolino introduces the subject matter of the book to the audience. This chapter starts with the definition of sensors and discusses different types of sensors and data analytics. The second chapter, Bio- and Nanosensors in the Food Industry by Charles Oluwaseun Adetunji, Frank Abimbola Ogundolie, Modupe Doris Ajiboye, John Tsado Mathew, Abel Inobeme, Olotu Titilayo, Olugbemi Tope Olaniyan, Oluwatosin Ademola Ijabadeniyi, Olulope Olufemi Ajayi, Wadazani Dauda and Shakira Ghazanfar, presents more examples of such novel applications related to the preservation, authenticity and safety of food items. The third chapter, Bio-based Sensing: Role of Natural Dyes in Food Freshness Indicators by V. G. Martins, L. G. Santos, V. P. Romani and S. S. Fernandes discusses the different kinds of natural dyes and their applications in food packaging. Chapter 4, Organic–inorganic Hybrid Materials for Active Packaging Applications by P. Wojciechowska presents different methods of synthesis of organic–inorganic hybrid materials and highlights their role for active packaging with examples. Chapter 5, Bio-based and Nanostructured Hybrids for Green and Active Food Packaging by Maria-Nefeli Efthymiou, Erminta Tsouko, Eleni Vlassi, Aristeidis Papagiannopoulos, Apostolis Koutinas and Stergios Pispas presents state-of-the-art applications of natural, bio-based and biodegradable polymeric materials in food packaging. Chapter 6, Smart Food

Food Chemistry, Function and Analysis
Bio- and Nano-sensing Technologies for Food Processing and Packaging
Edited by Ashutosh Kumar Shukla

Published by the Royal Society of Chemistry, www.rsc.org

Sensing and IoT Technologies by B. T. W. Putra, and B. Kuswandi presents IoT and artificial intelligence applications for smart sensing. Chapter 7, Nano-engineered Sensors for Food Processing by Charles Oluwaseun Adetunji, Frank Abimbola Ogundolie, Modupe Doris Ajiboye, John Tsado Mathew, Abel Inobeme, Wadazani Dauda, Shakira Ghazanfar, Olotu Titilayo, Olugbemi Tope Olaniyan, Oluwatosin Ademola Ijabadeniyi and Olulope Olufemi Ajayi illustrates the nano-engineered sensor applications with examples. Chapter 8, Nanobiomaterials for Food Packaging Sensor Applications by Charles Oluwaseun Adetunji, Olugbemi Tope Olaniyan, Frank Abimbola Ogundolie, John Tsado Mathew, Abel Inobeme, Olotu Titilayo, Shakira Ghazanfar, Oluwatosin Ademola Ijabadeniyi, Modupe Doris Ajiboye, Olulope Olufemi Ajayi and Wadazani Dauda highlights the applications of biological nanoparticle-based sensors. Chapter 9, Nanotechnology to Detect the Microbial Toxins in Stored Food by Amrita Bhanja, Reetuparna Nanda and Monalisa Mishra presents the applications of nanomaterials for detecting microbial toxins. Chapter 10, Precautionary Measures for Developing Nanosensors for the Food Industry by Sharda Sundaram Sanjay highlights the precautionary measures by summarizing the pros and cons of nanosensor applications for the food industry. The contributors have given their best efforts to bring the sense of completeness to the individual chapters.

I sincerely thank Nicki Dennis, Commissioning Editor, Food Science and Chemistry, at the Royal Society of Chemistry for giving me the opportunity to present this volume to its audience. I also wish to thank Liv Towers, Editorial Assistant, Books, for her support during the different stages. Finally, my grateful thanks go to the expert authors and reviewers for their help.

Ashutosh Kumar Shukla
Prayagraj, India

Contents

Food Chemistry, Function and Analysis
Bio- and Nano-sensing Technologies for Food Processing and Packaging
Edited by Ashutosh Kumar Shukla

Published by the Royal Society of Chemistry, www.rsc.org

CHAPTER 1

Sensors for the Food Industry: An Introduction

D. COZZOLINO

The University of Queensland, Queensland Alliance for Agriculture and Food Innovation, Centre for Nutrition and Food Sciences, St. Lucia Campus, Brisbane QLD 4072, Australia
Email: d.cozzolino@uq.edu.au

1.1 Introduction

The term sensor refers to a single or series of instruments that are based on the utilization of chemical and physical principles such as electrochemistry, fluorescence, thermal, surface plasmon resonance, piezo, reflectometry, chemo- or bioluminescence, and optics (*e.g.* vibrational spectroscopy).[1–13] More specifically, the term chemical sensor refers to or describes a small appliance that is able to deliver instant information about the chemicals and other compounds present in a food sample or food system that can be used throughout the different steps of the supply and value chain, including the processing, storage and marketing (*e.g.* on-line) of food ingredients and products.[1–13]

Current advances in sensing technologies have allowed for the development of novel applications in a wide range of fields, including food sciences.[1–13] Some of these advances that have boosted the implementation and utilization of these devices in food analysis include instrument miniaturization, the use of low sample volumes, improvements in sample integrity over sample destruction (non-destructive), advances in chemical recognition, ease of manufacture using recognition methods such as

Food Chemistry, Function and Analysis
Bio- and Nano-sensing Technologies for Food Processing and Packaging
Edited by Ashutosh Kumar Shukla

Published by the Royal Society of Chemistry, www.rsc.org

molecularly imprinted polymers (MIPS), automation, and remote sensing (*e.g.* at/in/on-line).[1–13] These developments have paved the way for developments in many low-cost and autonomous sensors that can be economically produced and deployed in a wide range of food applications.[1–13] One advantage of the use of these sensors is that they can continuously collect and report data, providing information and knowledge about the sample (*e.g.* chemical changes during manufacturing, spoilage) or system.[1–13] This has resulted in an unprecedented amount of real-time and continuous measurements that can be collected and implemented during the processing, storage and marketing of food and food ingredients allowing a better understanding of the production system, including the supply and value chain.[1–13] However, the utilization of these new technologies has presented new challenges, such as how information can be extracted from the application of sensing technologies to evaluate complex data sets.[14,15] These challenges can be overcome through the application of significantly more advanced data analytical techniques to accurately interpret the data collected and delivered by applying efficient management systems along the supply and value chains.[14,15]

Food safety and security have occurred as gradually significant public concerns appeared worldwide due to sub-quality foods being linked to increased food security issues (*e.g.* fraud, contamination), mortality, human distress, well-being, and economic burden.[16–19] Consequently, in the so-called digital economy where consumer awareness and expectations of safety are high, the food industry must comply with and meet the needs of consumers as well as food quality control guidelines and standards imposed by government and food authorities.[16–19] Consumers will also use this information to make informed purchase decisions about their preferences for specific foods and/or food ingredients with high quality and reasonable prices, and at the same time, must maintain high-quality standards and assurance of product safety.[16–19]

Fitting the end-user conformity with regulatory guidelines on food quality, sensing technologies designers and the food industry itself have responded with continuous improvements and developments of analytical methodologies and instrumental techniques.[16–19] For example, analytical methods such as liquid chromatography (LC), gas chromatography (GC) and high liquid performance chromatography (HPLC) have gained great importance in the analysis of foods and have been used in a wide number of applications in the field of food analytics.[16–19,21,26] However, chromatographic analysis is constrained by the rigidities of often elaborate sample preparation steps, homogenization, extraction and clean up before the analytical component can be tested and quantified.[16–19] These processing steps are often repeated multiple times, as many samples are needed to give an accurate result due to the number of interferences in the food matrix (*e.g.* food extracts, freeze dry material), which can result in inaccurate identification and false positives.[16–19] Issues such as widespread set-up and extraction, and clean-up processes required during HPLC analysis can also hamper the analysis of foods in issues such as contamination or fraud monitoring.[16–19] The cumbersome time requirements imposed during the use of HPLC and other instrumental analytical techniques make them unsuitable to be utilized to evaluate and monitor fresh food ingredients and products due to their intrinsic short shelf-life. Therefore, sensing technologies devices and platforms may present a viable alternative to the current instrumental methods used by the food industry.[16–19]

Concomitantly, public health and consumer safety authorities are demanding the development of rapid screening techniques that can be used in both the field and laboratory to evaluate not only the composition and quality of foods but also to guarantee their safety. These techniques should be relatively inexpensive, easy to operate and require little or no sample preparation, allowing them to be used in-line or at-line. Sensing techniques have demonstrated their potential to characterize the complexity of food and food ingredients rather than relying solely on subjective analysis by humans.[20–25] It has been defined by several scientists and researchers that the ideal instrumental method must be objective, cost-effective and provide rapid, reproducible results, with continuous operation.[20–25] However, up to now, sensing technologies have not been extensively used by the food industry due to the lack of training and knowledge about their capability as an analytical method, in addition to issues associated with their reliability that is highly dependent on the food matrix being analysed.[20–25]

This chapter offers a summary of the main characteristics of the most frequently used and novel sensors applied by the food industry to measure and monitor issues related to food contamination and spoilage, aroma and taste, among other applications.

1.2 Type of Sensors

1.2.1 Biosensors

A biosensor is defined as an analytical tool that combines a biologically sensitive recognition component (*e.g.* antibodies, nucleic acids, enzymes, whole cells and aptamers) restrained on a physicochemical transducer, and connected to a detector that allows the identification and quantification (*e.g.* concentration) of a single or several specific analytes in a food.[4,27–33] For example, electrochemical biosensors utilize an electrode transducer to identify electrons released by the reaction of the bioreceptor and analyte to acquire a measurable analysis of the contaminant present in a food.[4,21,27–36] One of the most important achievements of biosensors is that they can recognize biological elements, which provides them with a superior level of specificity and binding affinity with the target molecule.[35–40] These binding properties are termed specific binding or coupling and determine whether an interaction occurs, which creates the electrical signal that is recorded and amplified by the sensing technology during the analysis of food.[35–40] Because of the particularity of the recognition entity towards the analyte, a high level of selectivity is achieved which results in signals generated solely from such precise interactions, irrespective of the matrix complexity.[35–40]

Food security and safety demand high-speed, accurate and selective identification of the analytes present within the food matrix (*e.g.* pathogenic bacteria, presence of pesticides and toxins).[40–44] Developments in biosensors have enabled the quantitative detection and screening of different microorganisms and molecules in food. In addition to their inherent specificity, biosensors have demonstrated that they can overcome the multifactorial food industry challenges of high analytical accuracy in complex food matrices, overcoming issues such as differences in density, pH and temperature.[40–43] Another advantage of biosensors is that they have the potential for miniaturization, which results in the sensor requiring greatly reduced sample sizes or volumes.[40–43] However, miniaturization has only partially been achieved due to restrictions in the structural integrity of very fine electrode tips associated with microelectrodes.[40–43]

Recent developments in novel bio-recognition molecules, such as synthetic aptamers, DNA, proteins and viruses have increased the selectivity of molecules during analysis.[45–47] Improvements in the immobilization of bio-recognition molecules because of robust attachment methods like electrodeposition and nanoparticle-bound entities at the working electrode interface are a significant step in the increased application of biosensors in food analysis.[45–47] This is due in no small part to their greater specificity, selectivity and affinity for their target analytes.[45–47] The utilization of biosensors in the field of food analysis has grown over the last two decades as a result of improved accuracy in target detection, intensifying demands about quality and safety from stakeholders, such as safety regulators, traders and consumers as well as a significant reduction in the analysis times associated with electrochemical detection.[45–47] Enzymatic-based biosensors have been recognized as being of high selectivity, where the next stage in biosensor design includes gene-based sensors involving DNA, such as the recognition or coupling entity (*via* hybridization), antibody- or antigen-based biosensors, as well as whole-cell sensors.[45–47]

Within the agri-food industry, pathogen detection trends have focused on the utilization of a single-sensor platform for the detection of multiple pathogens/toxins.[48–50] More recently, biotechnology has shifted into ever smaller systems to allow for portability, cost reduction, analysis time reduction and commercial viability of this type of sensor.[48–50] Improvements in microfabrication systems have similarly aided in advancing biosensor technology and utility.[48–50]

Emerging nanomaterials, such as nanoparticles and nanofibres, play an important role in paving the way for this miniaturization trend.[51–60] These functional nanomaterials improve electrochemical biosensors by refining the response features of the electrode, increasing its surface area and assisting in robust attachment of the bioreceptor or recognition unit.[51–60] With a greater surface area to volume ratio, nanomaterials lend greater catalytic ability, warrant biocompatibility and achieve lower mass transfer resistance.

These properties improve the selectivity, sensitivity, time efficiency and cost-effectiveness of the biosensor.[51–60] Similarly, the increase in transducer surface area delivers greater conductivity and sensitivity, promoting greater interaction capacity, and lowering the detection limits.[51–60]

A recent example of a nano-biosensor is the detection of pesticide residues at low concentrations (*e.g.* below 0.4 pM).[51] The inclusion of other nanomaterials at the transducer level (*e.g.* carbon nanotubes) can increase electron transfer and increase the transducer activity.[51–60] The evidence of these improvements is in the slow but gradual replacement of traditional enzyme–substrate biosensors by nano-biosensor technology.[51–60] Nano-biosensors have been developed for the agriculture and food-processing industries to identify and quantify pesticides, herbicides, pathogenic microorganisms and other microbial contaminants or unwanted microorganisms such as viruses and bacteria, insects and fungi, as well as hormones and other foreign substances.[51–60]

Microfluidics have been also incorporated into biosensors as they can be utilized throughput processing, reducing sample and reagent volumes (down to the nanolitre), increase sensitivity, and employ a single platform for both sample preparation and detection.[61–64] The main advantage of microfluidics is that they can be portable, disposable and offer real-time detection, allowing for the simultaneous analysis of different analytes in a single device with exceptional accuracy.[61–64] The detection of pathogenic species present in food such as *Salmonella* is an example of the application of microfluidic nano-biosensors.[61–64]

1.2.2 Optical Sensors: Vibrational Spectroscopy

Existing techniques and methods used during the routine analysis of foods are founded on the determination and evaluation of the chemical composition of a given sample based on the analysis of a single parameter. From this, either the overall quality or the changes that have occurred during food processing and/or storage are inferred.[65–72] Current innovations in instrumentation (*e.g.* handheld, portable, hyperspectral, multispectral), computing and software are enhancing the development of novel tools and applications.[65–72] These technological advances are determining a fundamental change in the direction of the analytical methods currently in use and on the underlying assumptions of how food ingredients and products are evaluated.[65–72] These fundamental changes are accentuated by the current challenges facing the food industry such as food security, traceability, safety of food ingredients, prediction of functionality and consumer choices, that are generally difficult to measure using the current analytical approaches used by the food industry.[65–72] This has determined that the quantification of parameters associated with food quality is beyond the measurement of a single analyte, where the evaluation and monitoring of issues associated with food security and sustainability require changes in the way that food ingredients and products are analysed, to assure consumers about all aspects of food safety and security.[65–72]

It is not surprising that the amalgamation of sensing technologies (*e.g.* NIR spectroscopy) with data analytics is driving these current changes in the food industry towards so-called smart and sustainable food systems.[65–72] The incorporation of sensing technologies, the Internet of Things and Big Data has also been proven to reduce waste and losses during the transport and storage of food ingredients and products along the value chain. An example of this is the incorporation of wireless sensing devices that enable monitoring of fluctuations in both temperature and humidity during the transport and storage of foods.[73]

The capability of sensing technologies based on the utilization of vibrational spectroscopy to measure various parameters simultaneously, along with the speed of analysis, has demonstrated the huge potential value of these methods for incorporation in a sustainable food industry. These characteristics are also influential in the development of new approaches to food analysis and safety monitoring.[65–72] Progress in these types of applications is allowing the food industry to incorporate novel analytical methods that can be used to evaluate the composition of food as well as to monitor or detect changes (*e.g.* unwanted problems) providing rapid means of qualitative rather than quantitative analysis.[65–72]

Vibrational spectroscopic techniques have been developed to determine and measure biochemical, chemical and functional properties in different biological samples, including food ingredients and products.[65–72] Vibrational spectroscopy is the shared term utilized to describe the family of different techniques that include near-infrared (NIR), mid-infrared (MIR), Raman, terahertz spectroscopies and hyperspectral imaging.[65–72] Once food is analysed by these techniques, the vibrational spectrum behaves as the so-called "fingerprint" of the sample, providing the means to quantify, qualify, characterize and elucidate the intrinsic characteristics of food ingredients and products.[65–74] The main characteristics of vibrational spectroscopy are associated with the simple steps required for sample preparation, the non-destructive and rapid nature of the analysis, and it has been extensively used to analyse a wide range of food ingredients and products qualitatively and quantitatively.[65–78]

Vibrational spectroscopy (*e.g.* NIR, MIR and Raman) was first embraced by the agricultural and food sector as an analytical technique to predict the proximate composition of foods and raw ingredients in the early 1970s.[65–78] In vibrational spectroscopy, bonds such as C–H, O–H and N–H in the sample have high absorbance in both NIR and MIR regions. For example, the simple sample preparation required during the analysis of food samples in NIR spectroscopy makes it more feasible to monitor samples during processing (*e.g.* on-, in- and at-line), quality control and/or farm applications (*e.g.* from farm to fork).[65–78] On the other hand, MIR spectroscopy might require some level of sample preparation and can be influenced by sample moisture, however, it shows better specificity and reproducibility.[65–78] Therefore, the choice between NIR and MIR spectroscopy depends on the purpose of the specific application. Raman is often considered to be a complementary technique to both NIR and MIR spectroscopy.[65–78] Raman spectroscopy generates intense bands of functional groups with high polarizability (C=Cl, C=C, and C=N) in contrast to the strong absorbance bands of functional groups with strong polarization (O–H and C=O) in the infrared region.[78,79] The main bands in the spectra are due to local vibrational modes that emerge strongly in the infrared spectra, while stretching modes appear intensively in Raman spectra.[78,79] Raman spectroscopy is based on inelastic scattering from the interaction of incident radiations with vibrating molecules, where the main issue encountered in the application of this technique is its low sensitivity due to weak Raman scattering.[78,79] A wide range of techniques have been developed to overcome and improve the Raman signal such as stimulated Raman scattering, coherent anti-Stokes Raman scattering, resonance Raman spectroscopy and surface-enhanced Raman spectroscopy (SERS).[78,79]

Terahertz waves lie between the end of the IR region and the start of the microwave region.[80] The application of terahertz in the analysis of biological samples is still at an early stage.[80] The main advantage of THz is that this technique can be useful as a potential tool to characterize far-infrared vibrational modes in food with the main objective of water measurement. This is because THz waves coincide with the vibration of the hydrogen bond and it is vested with high sensitivity to water molecules.[80] However, this is also the limitation for applying THz spectroscopy to evaluate other compounds in biological samples with high moisture contents.[80] Therefore, sensors based on the utilization of vibrational spectroscopy can only provide point-based spectral information, and thus no spatial information for a large area will be available, although spatial information will be of especial interest in some biological applications.[80]

Hyperspectral imaging techniques can be considered a blend of the main characteristics of vibrational spectroscopy and imaging technique into one system, providing both spatial and spectral information about the sample.[81–86] During the implementation of this technique a hyperspectral image is constructed by a hypercube with a three-dimensional data set consisting of one spectral and two spatial dimensions.[81–86] Hyperspectral reflectance imaging is often accomplished by combining the visible and NIR wavelength ranges and has been widely applied to quantify different attributes, whereas Raman or Terahertz spectroscopy can also be integrated with imaging technology to form Raman or THz imaging.[81–86] Raman imaging is sensitive to specific chemicals at low concentrations, enabling visualization of the chemical distribution in samples where THz imaging might be an alternative to X-rays.[81–86]

1.2.3 Electronic Noses

The ultimate analytical tool to measure and monitor food aroma and flavour can be defined as such a technique that can quantify changes in the temporal dimension that are directly associated with perception.[87–94] To realize this in a timely manner and at low cost, analytical systems that provide high selectivity and sensitivity comparable with the human olfactory receptor are required during the analysis of foods.[87–94] Although techniques such as electroencephalography (EEG) and magnetic resonance imaging (MRI) can be used to monitor participants' brain activity directly associated with a taste or aroma stimulus, these instruments are not practical for application in industrial settings (*e.g.* due to excessive cost, lack of portability, time consuming).[87–97]

Electronic noses (EN) are instrumental devices that have been developed to mimic the perception of the mammalian olfactory system.[13] The definition and concept of EN is often associated with the recognition of odours (*e.g.* volatile compounds) or an attempt to mimic the smell process with an instrumental device.[13] The EN instruments offer the capability to detect different volatiles or gases with no odour activity. These instruments can be adapted to monitor or measure substances of importance to humans and animals, such as the scent of other animals, food ingredients or spoilage.[13]

A typical EN instrument consists of a series of an array of sensors.[87–94] This is the most common type of EN, although new technologies have been evaluated and reported by several authors such as metal oxide sensors (MOX),[95] mass spectrometry (MS), ion mobility spectrometry (IMS) and gas chromatography (GC)-based sensors.[13,20,22]

The utilization of EN has been evaluated by many researchers as a means to differentiate food and food ingredients based on the aroma and quantification of volatile compounds.[87] Food ingredients and products characterization is based on the analysis and detection of aroma properties.[87] For example, analytical solutions for food and wine composition often involve the use of GC-MS techniques, but analysis can be time consuming, due to sample preparation steps and complex data interpretation.[98,99] Electronic noses based on MS can detect mass fragments formed during the ionization of volatile compounds where some of these volatiles can be directly responsible for explaining or monitoring changes in the sensory properties of the samples.[98,99] Overall, EN instruments have found a niche application in the detection of food spoilage. It is well known that when food is stored, it releases specific volatile compounds, such as hydrocarbons, ethanol, aldehydes, acids, ethers and esters.[98,99] The presence and concentrations of these volatile compounds are different and depend, among other factors, on the raw material, brand and type of food, storage conditions (*e.g.* temperature, humidity) and issues associated with spoilage (*e.g.* the presence of microorganisms).[98–100] These volatile compounds have been proposed as fingerprints of the factors affecting food during storage and processing.[98–100]

Conducting organic polymers (COP) have been also explored as EN.[101–103] Instruments based on COP consist of a series of active mechanisms which detect odours and convert chemical vapours into electrical signals.[101–103] Conducting polymer can be affected by the doping levels, the ion size of the dopant, the water content and protonation levels. These types of sensors can be also classified depending on their mode of transduction and/or the application.[101–103]

Metal oxide semiconductors (MOS) also were evaluated as EN. Instruments based on MOS consist of a gas sensor which can detect changes in the concentration of volatile compounds in the sample as a function of a treatment, spoilage, *etc.*[102] Real-time quantitative detection of volatiles can be achieved using an electron impact source with a membrane separator between the source and the external environment coupled with a mass spectrometer.[88] The advantages of MS (relative to the electronic nose) are its fast response rate and greater sensitivity.[88,102] However, often the separator membrane reduces the technique's overall sensitivity, while its selective permeability reduces its application potential.[88,102]

More recently, further developments in instruments based on mass spectrometry systems, such as atmospheric pressure chemical ionization mass spectrometry (APCI-MS) and proton transfer reaction mass spectrometry (PTR-MS) are envisioned as being capable of the real-time detection of compounds at ppb concentrations.[22,88]

The utilization of EN devices has allowed unique volatile compounds to be distinguished within a wide range of food ingredients, beverages and products.[22,88,104,105] The application of EN provides a certain specificity, mainly due to the fact that one sensor might respond to more specific compounds such as an ester than aldehydes.[22,88,104,105] Unfortunately, none of the EN and sensor arrays available can respond in a manner comparable with that of the human olfactory system.[22,88,104,105] Furthermore, they lack both sensitivity (ppm/ppb concentration) and, in some cases, selectivity.[22,88,104,105] For example, EN coupled with their comparatively slow response rate make existing instrumentation incapable of appropriately monitoring rapid changes in breath volatile concentration.[22,88,104,105]

1.2.4 Electronic Tongue

Another sensing technology widely used is the electronic tongue. This instrument was designed in combination with biosensors that utilized electrochemical transducers to function.[106–110] Several types of chemical sensors have been used or design as electronic tongue instruments, such as a sensor array: electrochemical (potentiometric, voltammetric, amperometric, impedimetric, conductimetric), optical, mass and enzymatic biosensors.[106–110] An electronic tongue can be defined as an advanced analytical instrument made up of an array of sensors combined with pattern recognition technologies that are able to execute quantitative and qualitative analyses.[106–110] A full detailed characterization of the properties of the different electronic tongue instruments can be found elsewhere, where the most utilized are the potentiometric and voltammetric electronic tongue instruments.[106–110] It is well recognized that potentiometric electronic tongues are the most widely used, while voltammetric instruments have been proved to be more adaptable with high resistance to electrical imbalances and better signal-to-noise ratio.[106–110] Voltammetric electronic tongues are limited to redox-active substances and are sometimes associated with low detection limits.[106–110] The potentiometric electronic tongues have achieved notoriety in food quality assurance.[106–110] This type of electronic nose is made up of an array of non-selective chemical sensors with partial specificity to different constituents in solution.[106–110] Similarly to EN, signals from the sensors are interpreted using pattern recognition techniques that provide the potential to quantify or qualify the composition of a food.[106–110] Depending on the sensitivity and specificity, different materials have been used to build electronic noses, particularly if a high performance is desired.[106–110] Although electronic noses do not require special interactions with the analyte, the sensing unit is still required to respond electrically to small changes in the liquid under analysis.[106–110] However, depending on the method of detection, a level of electrical conductivity and/or electroactivity may be required.[106–110] These materials include lipid membranes, chalcogenide glasses, ion-sensitive field effect transistors (ISFET) and metal oxide semiconductor field effect transistors (MOSFET).[106–110]

1.2.5 Other Types of Sensors Available

Many other sensors have been developed or explored, such as colorimetric, a wide variety of electrochemical and chemical sensors, smartphone devices, among others. Please note that the in-depth characterization of the different sensors available is beyond the scope of this chapter, and it is suggested that the reader seek information from the vast number of publications, scientific and industry reports, and patents available in the field. Figure 1.1 schematically represents the applications and uses of sensors in food sciences (research and industry).

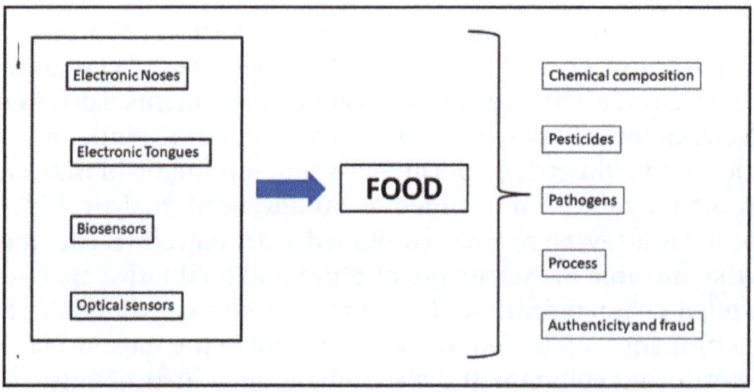

Figure 1.1 Applications and uses of sensors in food sciences (research and industry).

1.3 Data Analytics

One important aspect or characteristic of the use of sensors is the need for the integration with mathematical and statistical methods that must be required to analyse and interpret the data collected during analysis.[111–118] The utilization of both data analytics and chemometrics incorporates an extensive range of mathematical methods and algorithms that can be applied to analyse and optimize the data collected.[111–118] These methods are the result of the utilization solely or in combination of both qualitative and quantitative models and can often identify underlying relationships that exist between environmental variables, which cannot be independently identified using simple, univariate procedures.[111–118]

Several data analytics methods and techniques have been shown to be useful in the mining of data that have been obtained during the application and implementation of different types of sensors.[111–118] Some examples include the data mining of underlying factors that can be used to identify or monitor food contamination and spoilage, prediction of volatiles or other chemicals relevant to the food ingredient or food, to monitor the process, among others.[111–118] Most of the methods and techniques employed to extract information from these instrumental techniques have become part of so-called artificial intelligence (AI) or machine learning (ML) tools.[111–118] Machine learning tools are associated with the utilization of statistical methods to identify patterns in a data set and can be classified as unsupervised and supervised approaches.[111–118] These techniques allow the user to isolate and interpret information from complex data sets, illustrate patterns within the data, as well as the development of mathematical relationships between different sources of data (*e.g.* calibration and validation models).[111–118] Some of the most common and well-known techniques are principal component analysis (PCA) cluster analysis, K-nearest neighbours (k-NN), linear discriminant analysis (LDA), factorial discriminate analysis (FDA), partial least squares discriminant analysis (PLS-DA), quadratic discriminant analysis (QDA), artificial neural networks (ANN), soft independent modelling of class analogy (SIMCA), and more recently the group of machine learning methods and techniques [*e.g.* support vector machines (SVM)].[111–118]

1.4 Summary and Final Considerations

It is well accepted that sensing technologies have clear advantages over traditional analytical methods, however, the perfect sensing technique does not yet exist as there are many hindrances in its development and translation into real-world applications. Some of these issues are mostly associated with the implementation and commercial availability of these types of instrumentation. Nevertheless, it is inevitable that the future of sensing technologies will involve partnership with information communications technology to assist food producers, retailers, authorities and even consumers, in their decision-making process, preparing them with the necessary tools and data to improve their decision-making process. In addition, the implementation and utilization of sensing technologies will enable and improve our understanding of food systems, allowing for better management of natural resources, making food production more sustainable.

Networks of sensors will influence the development of new sensing systems that have the potential to revolutionize food analysis. These systems (*e.g.* single sensors or in combination) will be able to detect multiple signals allowing for a more holistic analysis of food quality and safety, and therefore the whole supply and value chain. The utilization of sensing technologies will also improve the ability and efficiency of current analytical systems, allowing for the evaluation and detection of multiple analytes simultaneously.

There is no doubt that the integration and implementation of different types of sensors have great promise in the field of food analytics. For example, it has been proposed that the fusion of electronic tongues with electronic noses could have a great impact as they increase the identification capabilities of these monitoring systems, coming close to mimicking the biological system (*e.g.* food). In addition, advantages such as real-time monitoring throughout food manufacture (*e.g.* dairy, brewing) further enhance the usefulness of sensing technologies driving the push for their commercial availability to the consumer. The intrinsic specificity, sensitivity and adaptability of most of the sensing technologies make them the ideal tool for use as a safety net throughout the food industry, improving product quality with minimal investment. The opportunity afforded through the utilization of sensing technologies, particularly *in situ* and safety analysis at all levels of the supply chain, as well as authenticity and quality analysis by consumers themselves, make food production sensors tools of the future.

The development and implementation of sensing technologies often requires the creation of mathematical models that can be used to quantify or monitor composition, contamination, fraud, *etc.* However, the importance of mathematical modelling is somehow ignored or underestimated. Sensing technologies have demonstrated that they can contribute to or can be used as tools in sustainable food systems, contributing to reducing the cost and time of analysis. As these techniques are reagent-less and have low energy consumption, they are defined as environmentally friendly.

Despite the huge potential of sensing technologies as analytical methods, their use by the food industry will require further developments in R&D. The current global demands for fully nutritious, sustainable and safe foods are on top of the agenda for the modern food industry and, more importantly, the consumer. Regrettably, these questions are hampered by several overarching issues, including rising complexity in the supply chains, the effects of climate change (*e.g.* composition of raw materials and ingredients), a growing ageing population, security issues (*e.g.* contamination, fraud, traceability, origin, waste) and continuous changes in consumer patterns or choices for sustainable and safe food ingredients and products.

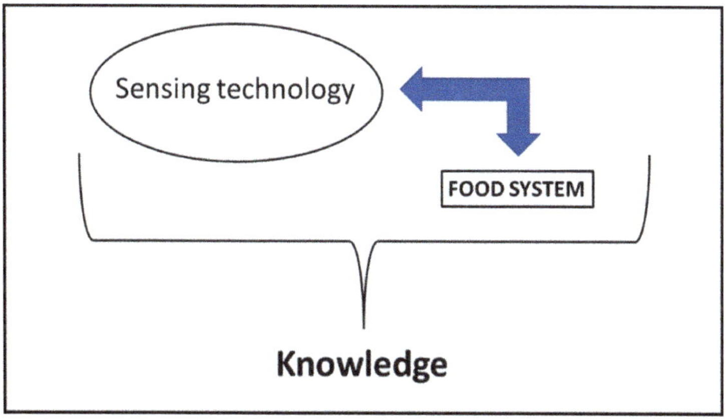

Figure 1.2 The utilization of sensing technologies and the creation of knowledge in the modern food industry.

Finally, the utilization of these sensing technologies will allow the extraction of knowledge derived from interrogation of the food system analysis (see Figure 1.2). This new approach will allow the development of novel solutions that will challenge as well as increase the efficiency, sustainability, flexibility, agility and resilience of the food supply and value chains, from the farmer to the consumer (*e.g.* a farm to fork approach).

Abbreviations

AI	Artificial intelligence
ANN	Artificial neural networks
APCI-MS	Atmospheric pressure chemical ionization mass spectrometry
COP	Conducting organic polymers
EN	Electronic nose
EEG	Electroencephalography
FDA	Factorial discriminate analysis
GC	Gas chromatography
GC-MS	Gas chromatography mass spectrometry
HPLC	High-performance liquid chromatography
IMS	Ion mobility spectrometry
ISFET	Ion-sensitive field effect transistors
k-NN	K-Nearest neighbours
LC	Liquid chromatography
LDA	Linear discriminant analysis
MIPS	Molecularly imprinted polymers
MIR	Mid infrared
ML	Machine learning
MOX	Metal oxide sensors

MOSFET	Metal oxide semiconductor field effect transistor
MRI	Magnetic resonance imaging
MS	Mass spectrometry
NIR	Near infrared
PCA	Principal component analysis
PLS-DA	Partial least squares discriminant analysis
PTR-MS	Proton transfer reaction mass spectrometry
QDA	Quadratic discriminant analysis
SERS	Surface-enhanced Raman spectroscopy
SVM	Support vector machines
SIMCA	Soft independent modelling of class analogy
THz	Terahertz

References

1. M. Y.-C. Wu, M.-Y. Hsu, S.-J. Chen, D.-K. Hwang, T.-H. Yen and C.-M. Cheng, Point-of-care detection devices for food safety monitoring: Proactive disease prevention, *Trends Biotechnol.*, 2017, **35**(4), 288–300.
2. F. Yeni, S. Acar, Ö. G. Polat, Y. Soyer and H. Alpas, Rapid and standardized methods for detection of foodborne pathogens and mycotoxins on fresh produce, *Food Control*, 2014, 40359–40367.
3. E. B. Bahadır and M. K. Sezgintürk, Applications of commercial biosensors in clinical, food, environmental, and biothreat/biowarfare analyses, *Anal. Biochem.*, 2015, 478107–478120.
4. V. Perumal and U. Hashim, Advances in biosensors: Principle, architecture and applications, *J. Appl. Biomed.*, 2014, **12**(1), 1–15.
5. J. Chapman, A. Power, K. Kiran and S. Chandra, New twists in the plot: Recent advances in electrochemical genosensors for disease screening, *J. Electrochem. Soc.*, 2017, **164**(13), B665–B673.
6. E. Korotkaya, Biosensors: Design, classification, and applications in the food industry, *Foods Raw Mater.*, 2014, **2**(2).
7. T. F. McGrath, C. T. Elliott and T. L. Fodey, Biosensors for the analysis of microbiological and chemical contaminants in food, *Anal. Bioanal. Chem.*, 2012, **403**(1), 75–92.
8. C. I. L. Justino, A. C. Freitas, R. Pereira, A. C. Duarte and T. A. P. Rocha Santos, Recent developments in recognition elements for chemical sensors and biosensors, *TrAC, Trends Anal. Chem.*, 2015, **68**, 2–17.
9. S. Ampuero and J. Bosset, The electronic nose applied to dairy products: a review, *Sens. Actuators, B*, 2003, **94**(1), 1–12.
10. P. N. Bartlett, J. M. Elliott and J. W. Gardner, Electronic noses and their application in the food industry, *Food Technol*, 1997, **51**, 44–48.
11. A. K. Deisingh, D. C. Stone and M. Thompson, Applications of electronic noses and tongues in food analysis, *Int. J. Food Sci. Technol.*, 2004, **39**(6), 587–604.

12. A. Loutfi, S. Coradeschi, G. K. Mani, P. Shankar and J. B. B. Rayappan, Electronic noses for food quality: A review, *J. Food Eng.*, 2015, **144**, 103–111.
13. F. Röck, N. Barsan and U. Weimar, Electronic nose: current status and future trends, *Chem. Rev.*, 2008, **108**(2), 705–725.
14. D. Cozzolino, Food for thought: the digital disruption and the future of food production, *Curr. Res. Nutr. Food Sci.*, 2019, 7, 607–609.
15. V. K. Truong, M. Dupont, A. Elbourne, S. Gangadoo, P. Rajapaksha Pathirannahalage, S. Cheeseman, J. Chapman and D. Cozzolino, From academia to reality check: a theoretical framework on the use of chemometric, *Foods*, 2019, **8**, 1–10.
16. S. Vigneshvar, C. C. Sudhakumari, B. Senthilkumaran and H. Prakash, Recent advances in biosensor technology for potential applications – an overview, *Front. Bioeng. Biotechnol.*, 2016, **4**, 11–18.
17. R. Pilolli, L. Monaci and A. Visconti, Advances in biosensor development based on integrating nanotechnology and applied to food-allergen management, *TrAC, Trends Anal. Chem.*, 2013, 4712–4726.
18. C. Qin, L. Tao, Y. H. Phang, C. Zhang, S. Y. Chen, P. Zhang, Y. Tan, Y. Y. Jiang and Y. Z. Chen, The assessment of the readiness of molecular biomarker-based mobile health technologies for healthcare applications, *Sci. Rep.*, 2015, **5**, 17854.
19. B. S. Sekhon, Nanotechnology in agri-food production: An overview, *Nanotechnol., Sci. Appl.*, 2014, 7, 31–53.
20. A. Guadarrama, J. A. Fernandez, M. Iñiguez, J. Souto and J. A. de Saja, Discrimination of wine aroma using an array of conducting polymer sensors in conjunction with solid-phase micro-extraction (SPME) technique, *Sens. Actuators*, 2001, 77, 401–408.
21. A. P. F. Turner, Biosensors: Sense and sensibility, *Chem. Soc. Rev.*, 2013, **42**, 3184–3196.
22. G.-J. E. Nychas, E. Z. Panagou and F. Mohareb, Novel approaches for food safety management and communication, *Curr. Opin. Food Sci.*, 2016, 1213–1220.
23. A. K. Deisingh, D. C. Stone and M. Thompson, Applications of electronic noses and tongues in food analysis, *Int. J. Food Sci. Technol.*, 2004, **39**, 587–604.
24. A. Gutierrez, J. A. Burgos, C. Garcera, A. I. Padilla, M. Zarzo, C. Chirivella, M. L. Ruiz and E. Molto, Optimization of an aroma sensor for assessing grape quality for wine making, *Span. J. Agric. Res.*, 2007, **5**, 157–163.
25. F. Röck, N. Barsan and U. Weimar, Electronic nose: current status and future trends, *Chem. Rev.*, 2008, **108**, 705–725.
26. F. Cacciola, P. Dugo and L. Mondello, Multidimensional liquid chromatography in food analysis, *TrAC, Trends Anal. Chem.*, 2017, **96**, 116–123.
27. D. R. Thévenot, K. Toth, R. A. Durst and G. S. Wilson, Electrochemical biosensors: Recommended definitions and classification, *Biosens. Bioelectron.*, 2001, **16**(1), 121–131.

28. E.-H. Yoo and S.-Y. Lee, Glucose biosensors: An overview of use in clinical practice, *Sensors*, 2010, **10**(5), 4558–4576.
29. I. Bazin, S. A. Tria, A. Hayat and J.-L. Marty, New biorecognition molecules in biosensors for the detection of toxins, *Biosens. Bioelectron.*, 2017, 87285–87298.
30. V. Gaudin, Advances in biosensor development for the screening of antibiotic residues in food products of animal origin – a comprehensive review, *Biosens. Bioelectron.*, 2017, 90363–90377.
31. V. Velusamy, K. Arshak, O. Korostynska, K. Oliwa and C. Adley, An overview of foodborne pathogen detection: In the perspective of biosensors, *Biotechnol. Adv.*, 2010, **28**(2), 232–254.
32. F. Bettazzi, G. Marrazza, M. Minunni, I. Palchetti and S. Scarano, Biosensors and related bioanalytical tools, *Compr. Anal. Chem.*, 2017, 77, 1–33.
33. J. Ali, J. Najeeb, M. A. Ali, M. F. Aslam and A. Raza, Biosensors: Their fundamentals, designs, types and most recent impactful applications: A review, *J. Biosens. Bioelectron.*, 2017, **8**, 235–244.
34. N. Bhalla, P. Jolly, N. Formisano and P. Estrela, Introduction to biosensors, *Essays Biochem.*, 2016, **60**(1), 1–15.
35. A. Sharma, K. Y. Goud, A. Hayat, S. Bhand and J. L. Marty, Recent advances in electrochemical-based sensing platforms for aflatoxins detection, *Chemosensors*, 2017, **5**, 1–15.
36. R. Singh, M. D. Mukherjee, G. Sumana, R. K. Gupta, S. Sood and B. D. Malhotra, Biosensors for pathogen detection: A smart approach towards clinical diagnosis, *Sens. Actuators, B*, 2014, 197385–197404.
37. L. Rotariu, F. Lagarde, N. Jaffrezic-Renault and C. Bala, Electrochemical biosensors for fast detection of food contaminants – trends and perspective, *TrAC, Trends Anal. Chem.*, 2016, 7980–7987.
38. U. Andjelković, M. Gavrović-Jankulović, T. Martinović and D. Josić, Omics methods as a tool for investigation of food allergies, *TrAC, Trends Anal. Chem.*, 2017, **96**, 105–117.
39. T. Lavecchia, A. Tibuzzi and M. T. Giardi, *Biosensors for functional food safety and analysis*, ed. M. T. Giardi, G. Rea and B. Berra, Springer US, Boston, MA, 2010, pp. 267–281.
40. C. C. Adley, Past, present and future of sensors in food production, *Foods*, 2014, **3**(3), 491–510.
41. D. S. Jayas, The role of sensors and bio-imaging in monitoring food quality, *Resour. Mag.*, 2017, 12.
42. N. A. Moreb, A. Priyadarshini and A. K. Jaiswal, Knowledge of food safety and food handling practices amongst food handlers in the republic of ireland, *Food Control*, 2017, 80341–80349.
43. M. S. Thakur and K. V. Ragavan, Biosensors in food processing, *J. Food Sci. Technol.*, 2013, **50**(4), 625–641.
44. P. Poltronieri, V. Mezzolla, E. Primiceri and G. Maruccio, Biosensors for the detection of food pathogens, *Foods*, 2014, **3**(3), 511–526.

45. Y. Zeng, Z. Zhu, D. Du and Y. Lin, Nanomaterial-based electrochemical biosensors for food safety, *J. Electroanal. Chem.*, 2016, 781147–781154.
46. C. Luo, Y. Lei, L. Yan, T. Yu, Q. Li, D. Zhang, S. Ding and H. Ju, A rapid and sensitive aptamer-based electrochemical biosensor for direct detection of escherichia coli O111, *Electroanalysis*, 2012, **24**(5), 1186–1191.
47. X. Ma, Y. Jiang, F. Jia, Y. Yu, J. Chen and Z. Wang, An aptamer-based electrochemical biosensor for the detection of salmonella, *J. Microbiol. Methods*, 2014, 9894–9898.
48. I.-H. Cho, A. D. Radadia, K. Farrokhzad, E. Ximenes, E. Bae, A. K. Singh, H. Oliver, M. Ladisch, A. Bhunia and B. Applegate, *et al.*, Nano/micro and spectroscopic approaches to food pathogen detection, *Annu. Rev. Anal. Chem.*, 2014, 7(1), 65–88.
49. K. Warriner, S. M. Reddy, A. Namvar and S. Neethirajan, Developments in nanoparticles for use in biosensors to assess food safety and quality, *Trends Food Sci. Technol.*, 2014, **40**(2), 183–199.
50. B. Derkus, Applying the miniaturization technologies for biosensor design, *Biosens. Bioelectron.*, 2016, 79901–79913.
51. M. L. Verma, Nanobiotechnology advances in enzymatic biosensors for the agri-food industry, *Environ. Chem. Lett.*, 2017, 1–6.
52. C. Zhu, G. Yang, H. Li, D. Du and Y. Lin, Electrochemical sensors and biosensors based on nanomaterials and nanostructures, *Anal. Chem.*, 2015, **87**(1), 230–249.
53. J. Anu Bhushani and C. Anandharamakrishnan, Electrospinning and electrospraying techniques: Potential food based applications, *Trends Food Sci. Technol.*, 2014, **38**(1), 21–33.
54. R. Eivazzadeh-Keihan, P. Pashazadeh, M. Hejazi, M. de la Guardia and A. Mokhtarzadeh, Recent advances in nanomaterial-mediated bio and immune sensors for detection of aflatoxin in food products, *TrAC, Trends Anal. Chem.*, 2017, 87112–87128.
55. M. Rai, P. S. Jogee and A. P. Ingle, Emerging nanotechnology for detection of mycotoxins in food and feed, *Int. J. Food Sci. Nutr.*, 2015, **66**(4), 363–370.
56. M. M. Barsan, M. E. Ghica and C. M. A. Brett, Electrochemical sensors and biosensors based on redox polymer/carbon nanotube modified electrodes: A review, *Anal. Chim. Acta*, 2015, 8811–8823.
57. N. Yang, X. Chen, T. Ren, P. Zhang and D. Yang, Carbon nanotube based biosensors, *Sens. Actuators, B*, 2015, 207690–207715.
58. T. K. Sharma, R. Ramanathan, R. Rakwal, G. K. Agrawal and V. Bansal, Moving forward in plant food safety and security through nanobiosensors: Adopt or adapt biomedical technologies? *Proteomics*, 2015, **15**(10), 1680–1692.
59. P. Pashazadeh, A. Mokhtarzadeh, M. Hasanzadeh, M. Hejazi, M. Hashemi and M. De La Guardia, Nano-materials for use in sensing

of salmonella infections: Recent advances, *Biosens. Bioelectron.*, 2017, 871050–871064.
60. R. Sharma, K. V. Ragavan, M. S. Thakur and K. S. M. S. Raghavarao, Recent advances in nanoparticle based aptasensors for food contaminants, *Biosens. Bioelectron.*, 2015, 74612–74627.
61. G. Luka, A. Ahmadi, H. Najjaran, E. Alocilja, M. DeRosa, K. Wolthers, A. Malki, H. Aziz, A. Althani and M. Hoorfar, Microfluidics integrated biosensors: A leading technology towards lab-on-a-chip and sensing applications, *Sensors*, 2015, **15**, 1–15.
62. D. G. Rackus, M. H. Shamsi and A. R. Wheeler, Electrochemistry, biosensors and microfluidics: A convergence of fields, *Chem. Soc. Rev.*, 2015, **44**(15), 5320–5340.
63. Y. Huang, Y. Shi, H. Y. Yang and Y. Ai, A novel single-layered mos2 nanosheet based microfluidic biosensor for ultrasensitive detection of DNA, *Nanoscale*, 2015, 7(6), 2245–2249.
64. G. Kim, J.-H. Moon, C.-Y. Moh and J.-G. Lim, A microfluidic nano-biosensor for the detection of pathogenic salmonella, *Biosens. Bioelectron.*, 2015, **67**(Supplement C), 243–247.
65. K. B. Bec, J. Grabska and C. W. Huck, Review near-infrared spectroscopy in bio-applications, *Molecules*, 2020, **25**, 2948.
66. K. B. Bec and C. W. Huck, Breakthrough potential in near-infrared spectroscopy: spectra simulation. A review of recent developments, *Front. Chem.*, 2019, **7**, 48–53.
67. T. M. P. Cattaneo and A. Stellari, Review: NIR spectroscopy as a suitable tool for the investigation of the horticultural field, *Agronomy*, 2019, **9**, 503.
68. C. Pasquini, Near infrared spectroscopy: A mature analytical technique with new perspectives—A review, *Anal. Chim. Acta*, 2018, **1026**, 8–36.
69. K. B. Walsh, V. A. McGlone and D. H. Hanc, The uses of near infra-red spectroscopy in postharvest decision support: A review, *Postharvest Biol. Technol.*, 2020, **163**, 111139.
70. D. Cozzolino and J. J. Roberts, Applications and developments on the use of vibrational spectroscopy imaging for the analysis, monitoring and characterisation of crops and plants, *Molecules*, 2016, **21**, 755–763.
71. B. M. Nicolai, K. Beullens, E. Bobelyn, A. Peirs, W. Saeys, K. I. Theron and J. Lammertyn, Non-destructive measurement of fruit and vegetable quality by means of NIR spectroscopy: a review, *Postharvest Biol. Technol.*, 2007, **46**, 99–118.
72. W. Saeys, N. N. Do Trong, R. Van Beers and B. M. Nicolai, Multivariate calibration of spectroscopic sensors for postharvest quality evaluation: A review, *Postharvest Biol. Nanotechnol.*, 2019, **158**, 110981.
73. R. A. Crocombe, Portable spectroscopy, *Appl. Spectrosc.*, 2018, **72**, 1701–1751.

74. K. Naseer, S. Ali and J. Qazi, ATR-FTIR spectroscopy as the future of diagnostics: a systematic review of the approach using bio-fluids, *Appl. Spectrosc. Rev.*, 2021, **56**(2), 85–97.
75. V. Cortes, J. Blasco, N. Aleixos, S. Cubero and P. Talensa, Monitoring strategies for quality control of agricultural products using visible and near-infrared spectroscopy: A review, *Trends Food Sci. Technol.*, 2019, **85**, 138–148.
76. R. Kaavya, R. Pandiselvam, M. Mohammed, R. Dakshayani, A. Kothakota, S. V. Ramesh, D. Cozzolino and C. Ashokkumar, Application of infrared spectroscopy techniques for assessment of quality and safety in spices – state-of-the-art and future trends spectroscopy techniques- an emerging tool for spices and herbs authentication, *Appl. Spectrosc. Rev.*, 2021, **55**, 593–611.
77. R. Karoui, G. Downey and Ch. Blecker, Mid-Infrared Spectroscopy Coupled with Chemometrics: A Tool for the Analysis of Intact Food Systems and the Exploration of Their Molecular Structure-Quality Relationships – A Review, *Chem. Rev.*, 2010, **110**, 6144–6168.
78. D. Sorak, L. Herberholz, S. Iwascek, S. Altinpinar, F. Pfeifer and H. W. Siesler, New developments and applications of handheld Raman, mid-infrared, and near infrared spectrometers, *Appl. Spectrosc. Rev.*, 2012, **47**, 83–115.
79. R. R. Jones, D. C. Hooper, L. Zhang, D. Wolverson and V. Valev, Raman Techniques: Fundamentals and Frontiers, *Nanoscale Res. Lett.*, 2019, **14**, 231.
80. L. Afsah-Hejri, P. Hajeb, P. Ara and R. J. Ehsan, A Comprehensive Review on Food Applications of Terahertz Spectroscopy and Imaging, *Compr. Rev. Food Sci. Food Saf.*, 2019, **18**, 1563.
81. M. L. Amodio, M. A. Chaudhry and G. Colelli, Spectral and hyperspectral technologies as an additional tool to increase information on quality and origin of horticultural crops, *Agronomy*, 2020, **10**, 7.
82. Y. Z. Feng and D. W. Sun, Application of hyperspectral imaging in food safety inspection and control: a review, *Crit. Rev. Food Sci. Nutr.*, 2012, **52**, 1039–1058.
83. M. Manley, Near-infrared spectroscopy and hyperspectral imaging: non-destructive analysis of biological materials, *Chem. Soc. Rev.*, 2014, **43**, 8600.
84. A. Baiano, Applications of hyperspectral imaging for quality assessment of liquid based and semi-liquid food products: a review, *J. Food Eng.*, 2017, **214**, 10–15.
85. D. Saha and A. Manickavasagan, Machine learning techniques for analysis of hyperspectral images to determine quality of food products: A review, *Curr. Res. Food Sci.*, 2021, **4**, 28–44.
86. J. M. Amigo, I. Martí and A. Gowen, Hyperspectral imaging and chemometrics: a perfect combination for the analysis of food structure, composition and quality, *Data Handl. Sci. Technol.*, 2013, **28**, 343–370.

87. P. N. Bartlett, J. E. Elliot and J. W. Gardner, Electronic noses and their application in the food industry, *Food Technol.*, 1997, **51**, 44–48.
88. R. S. T. Linforth, Developments in instrumental techniques for food flavour evaluation: future prospects, *J. Sci. Food Agric.*, 2000, **80**(14), 2044–2048.
89. C. F. Ross, Sensory science at the human–machine interface, *Trends Food Sci. Technol.*, 2009, **20**(2), 63–72.
90. E. Pionnier, C. Chabanet, L. Mioche, J.-L. Le Quéré and C. Salles, 1. *In vivo* aroma release during eating of a model cheese: relationships with oral parameters, *J. Agric. Food Chem.*, 2004, **52**(3), 557–564.
91. T. C. Pearce, S. S. Schiffman, H. T. Nagle and J. W. Gardner, *Handbook of Machine Olfaction: Electronic Nose Technology*, Wiley-VCH Verlag GmbH & Co. KGaA, 2006.
92. Z. Z. John Lewis, B. George, G. Zoltan and K. Zoltan, Emerging trends of advanced sensor based instruments for meat, poultry and fishquality–a review, *Crit. Rev. Food Sci. Nutr.*, 2020, **60**(20), 3443–3460.
93. M. P. Martí, O. Busto, J. Guasch and R. Boqué, Electronic noses in the quality control of alcoholic beverages, *TrAC, Trends Anal. Chem.*, 2005, **24**(1), 57–66.
94. P. Mielle, Electronic noses: Towards the objective instrumental characterization of food aroma, *Trends Food Sci. Technol.*, 1996, 7(12), 432–438.
95. A. Berna, S. Trowell, W. Cynkar and D. Cozzolino, Comparison of metal oxide based-electronic nose and mass spectrometry-based electronic nose to predict red wine spoilage, *J. Agric. Food Chem.*, 2008, **56**, 3238–3244.
96. A. Gliszczy, Electronic nose as a tool for monitoring the authenticity of food. A Review, *Food Anal. Methods*, 2017, **10**, 1800–1816.
97. T. Majchrzak, W. Wojnowski, T. Dymerski, J. Gębicki and J. Namieśnik, Electronic noses in classification and quality control of edible oils: A review, *Food Chem.*, 2018, **246**, 192–201.
98. M. Gallo and P. Ferranti, The evolution of analytical chemistry methods in foodomics, *J. Chromatogr. A*, 2016, **1428**, 3–15.
99. M. Ghasemi-Varnamkhasti and J. Lozano, Electronic nose as an innovative measurement system for the quality assurance and control of bakery products: a review, *Eng. Agric. Environ. Food*, 2016, **9**(4), 365–374.
100. S. Matindoust, M. Baghaei-Nejad, Z. Zou and L.-R. Zheng, Food quality and safety monitoring using gas sensor array in intelligent packaging, *Sens. Rev.*, 2016, **36**(2), 169–183.
101. A. Kukla, A. Pavluchenko, Y. M. Shirshov, N. Konoshchuk and O. Y. Posudievsky, Application of sensor arrays based on thin films of conducting polymers for chemical recognition of volatile organic solvents, *Sens. Actuators, B*, 2009, **135**(2), 541–551.
102. A. Sanaeifar, H. ZakiDizaji, A. Jafari and M. de la Guardia, Early detection of contamination and defect in foodstuffs by electronic nose: A review, *TrAC, Trends Anal. Chem.*, 2017, **97**, 257–271.

103. G. M. Suppes, B. A. Deore and M. S. Freund, Porous conducting polymer/heteropolyoxometalate hybrid material for electrochemical supercapacitor applications, *Langmuir*, 2008, **24**(3), 1064–1069.
104. A. Wilson and M. Baietto, Applications and advances in electronic-nose technologies, *Sensors*, 2009, **9**(7), 5099–5148.
105. D. M. Wilson, S. Hoyt, J. Janata, K. Booksh and L. Obando, Chemical sensors for portable, handheld field instruments, *IEEE Sens. J.*, 2001, **1**(4), 256–274.
106. L. Escuder-Gilabert and M. Peris, Review: Highlights in recent applications of electronic tongues in food analysis, *Anal. Chim. Acta*, 2010, **665**(1), 15–25.
107. A. Rita, D. Rosa, F. Leone, F. Cheli and V. Chiofalo, Fusion of electronic nose, electronic tongue and computer vision for animal source food authentication and quality assessment e A review, *J. Food Eng.*, 2017, **210**, 62–75.
108. F. Winquist, Voltammetric electronic tongues – Basic principles and applications, *Microchim. Acta*, 2008, **163**, 3–10.
109. M. Valle, A. Mimendia and J. M. Gutie, A review of the use of the potentiometric electronic tongue in the monitoring of environmental systems, *Environ. Modell. Software*, 2010, **25**, 1023–1030.
110. P. Ciosek and W. Wroblewski, Potentiometric electronic tongues for foodstuff and biosample recognition—An overview, *Sensors*, 2011, **11**(5), 4688–4701.
111. D. Cozzolino, The sample, the spectra and the maths – the critical pillars in the development of robust and sound vibrational spectroscopy applications, *Molecules*, 2020, **25**, 3674.
112. S. Bureau, D. Cozzolino and C. J. Clark, Contributions of Fourier-transform mid infrared (FT-MIR) spectroscopy to the study of fruit and vegetables: A review, *Postharvest Biol. Technol.*, 2019, **148**, 1–14.
113. L. Agelet and C. H. Hurburgh, A Tutorial on Near Infrared Spectroscopy and its' Calibration, *Crit. Rev. Anal. Chem.*, 2010, **40**, 246–260.
114. R. Bro and A. K. Smilde, Principal component analysis, *Anal. Methods*, 2014, **6**, 2812–2831.
115. E. Szymańska, J. Gerretzen, J. Engel, B. Geurts, L. Blanchet and L. M. Buydens, Chemometrics and qualitative analysis have a vibrant relationship, *TrAC, Trends Anal. Chem.*, 2015, **69**, 34–51.
116. E. Szymanska, Modern data science for analytical chemical data: a comprehensive review, *Anal. Chim. Acta*, 2018, **1028**, 1–10.
117. M. P. Callao and I. Ruisánchez, An overview of multivariate qualitative methods for food fraud detection, *Food Control*, 2018, **86**, 283–293.
118. J. P. Chavas and C. Nauges, Uncertainty, learning and technology adoption in agriculture, *Appl. Econ. Perspect. Policy*, 2020, **42**, 42–53.

CHAPTER 2

Bio- and Nanosensors in the Food Industry

CHARLES OLUWASEUN ADETUNJI,*[a]
FRANK ABIMBOLA OGUNDOLIE,[b] MODUPE DORIS AJIBOYE,[h]
JOHN TSADO MATHEW,[c] ABEL INOBEME,[d] OLOTU TITILAYO,[e]
OLUGBEMI TOPE OLANIYAN,[f]
OLUWATOSIN ADEMOLA IJABADENIYI,[g]
OLUOPE OLUFEMI AJAYI,[i] WADZANI PALNAM DAUDA,[j]
SHAKIRA GHAZANFAR[k] AND JULIANA BUNMI ADETUNJI[l]

[a] Applied Microbiology, Biotechnology and Nanotechnology Laboratory, Department of Microbiology, Edo University Iyamho, PMB 04, Auchi, Edo State, Nigeria; [b] Department of Biotechnology, Baze University, Abuja, Nigeria; [c] Plant Science and Biotechnology, Faculty of Science, Federal University, Oye Ekiti, Nigeria; [d] Department of Chemistry, Ibrahim Badamasi Babangida University Lapai, Niger State 9111014, Nigeria; [e] Department of Chemistry, Edo University, Iyamho, Nigeria; [f] Laboratory for Reproductive Biology and Developmental Programming, Department of Physiology, Rhema University, Aba, Nigeria; [g] Laboratory for Reproductive Biology and Developmental Programming, Department of Physiology, Edo University Iyamho, Nigeria; [h] Department of Biotechnology and Food Science, Durban University of Technology, Durban 4001, South Africa; [i] Department of Biochemistry, Edo State University Uzairue, formerly Edo University, Iyamho; [j] Crop Science Unit, Department of Agronomy, Federal University Gashua, Gashua P.M.B 1005, Yobe State, Nigeria; [k] Functional Genomics and Bioinformatics, National Agricultural Research Centre, Islamabad-45500, Pakistan; [l] Department of Biochemistry, Osun State University, Nigeria
*Emails: adetunjicharles@gmail.com; adetunji.charles@edouniversity.edu.ng

Food Chemistry, Function and Analysis
Bio- and Nano-sensing Technologies for Food Processing and Packaging
Edited by Ashutosh Kumar Shukla

Published by the Royal Society of Chemistry, www.rsc.org

2.1 Introduction

Nanotechnology has become an amazing technology that has revolutionized several industries, including the food industry. It is a field that focuses on materials such as atoms, molecules, and macromolecules that are at the nanoscale level. These particles have an approximate dimension within 1–100 nanometres. It has been documented that these materials possess features that differ from their counterparts in the macro-scale level due to their unique physicochemical qualities connected with their high surface area to volume ratio, diffusivity, solubility, optical and thermodynamic parameters.[1] As a result of the skyrocketing number of studies in the area of utilization of nanotechnology in various areas, many researchers have shifted their focus to the application of bio- and nanosensors in the food sector.[2–12] This has been necessitated due to the increasing concerns of consumers about food quality and its nutritional benefits with a view to enhancing the overall quality of food and feeding the fast-growing global population. Due to the fact that most nano-based products have been found to have essential nutrients and are nontoxic, there has been an increasing demand for their application in today's food sector.[13]

One of the most recent areas of application of nanotechnology in the food sector is nanosensing technology. Nano- and biosensors have been employed in the achievement of efficient food quality and reliable safety assessment and evaluation. Nanosensing technology has found relevance at various stages of food production such as manufacturing, processing and also the packaging of the final products. It not only brings about an improvement of food overall quality but also in the health benefits of foods. Efficient food packaging aids in the prevention of contamination, moisture and spoilage. The conventional approaches to packaging and processing have their limitations. Hence they have failed to address the rising concerns about the safe packaging of food. The need for smart food processing and packaging that is nano-technologically driven has become pressing. Smart packaging makes use of various kinds of materials such as biosensors, nanosensors, temperature and time indicators amongst others.[14] The application of nanosensors, as well as nano-packaging, enhances sensitive, rapid and reliable determination of contamination due to microorganisms, poisonous chemicals, as well as pesticide residues. Nanosensors are efficient in the detection of various substances such as microbes, toxins, vitamins, adulterants, pesticides, fertilizers, smell and taste. The freshness of food can also be effectively monitored.[15]

The application of biosensors and nanobiosensors has enabled the efficient monitoring and provision of reliable information on the quality of food materials. Such sensors can efficiently alert consumers in the presence of toxic chemicals, pathogenic contamination, *etc.* The packaging of food is a vital component of food processing as it helps in the extension of food shelf-life, maintenance of food quality, and improving the safety of food products. The integration of nanosensors into the area of food processing and packaging has brought about remarkable advantages in comparison to the traditional sensors due to their enhanced specificity, sensitivity, analysis speed and reliability.[16]

This review, therefore, focuses on the application of bio- and nanosensors in the food sector. It also provides background information on nanosensors in the food industry and highlights some of the major challenges associated with the application of this novel technology in the food sector. Finally, current trends in bio- and nanosensing technology in the food industry are also highlighted.

2.2 Nanosensors in the Food Industry

The rapid advancement of nanotechnology has created several more dimensions to food security scientific facts and also the food industry, resulting in increased market and investment share. Advanced technologies in nanodevices and nanomaterials, including nanoemulsions, nanosensors, nanocapsules as well as nanopesticides, are expected to result in the novel food industrial applications. The existing nanotechnology applications for processing, packaging, enhancing the nutritional benefit, and also storage stability of foods are the main focus of research. Furthermore, the applicability and functionality of nanotechnology food-related applications provide for an evaluation of nanotechnology as well as intuition regarding safety in the food service industry.[17]

Science and technology, particularly in the area of nanotechnology, are being used in food industries to ensure the safety of food, tangible and intangible food manufacturing and pollutant identification. The confirmation of food safety, as well as its quality, has emerged as a critical concern for governments and also for food supply-chain participants. Food analysis is absolutely essential to protect purchasers from adulteration, spoilage and contamination, as it can identify various chemicals that may be detrimental to humans and the environment. Food systems are growing more convoluted, ecological limitations become more stringent, and also consumers have been altering the way they purchase and eat food, which together have led to modern cultures being much more concerned about possible dangerous compounds that may be found in foods. The application of nano-based particles is very broad, with applications reported in several industries including the food manufacturing, hence additional research is needed. The prospects for both food sector to benefit through nanoscience should grow as advances in nanotechnology development and research continue.[18]

It has been discovered that these microorganisms have the capacity to utilize macronutrients create food poisons as a defensive measure. Pathogens contaminate food, alter its texture, and therefore can cause disease, which may result in severe ill health or even death. In the food business, brand protection, as well as the prevention of food adulteration, is also a key concern. Researchers look at how nanoparticles can be used to detect the quality of food. Bacterial species, food-contaminating chemicals, adulterants, vitamins, colours, pesticides, fertilizers, smells and taste can all be detected with nanosensors. Time and temperature, as well as oxygen metrics, can be used to measure food freshness. Invisible nanobarcodes can be used to ensure product brand protection as well as determine authenticity. Generally, food security is being improved through nanosensors that exhibit distinctive features.[19]

The advantages of using nano-based technologies in food packaging over traditional sensing include enhanced specificity and sensitivity, faster assessment, higher sample productivity (sequencing devices), lower assay complexities and lower cost. The application and constituents in the design of three major kinds of nanosensors are discussed in this chapter: gas sensing, nanobiosensors and radio-frequency identification. For a single substance, there may be a range of nanosensor designs available, each with a distinct level of specificity and sensitivity. Some potential sources of risk to the marketing of these technologies (for example, the diversity of food materials) also are addressed. Different materials with varied physical and chemical compositions may be mixed to enhance the performance using these sensors and eliminate a few of the constraints, in addition to developing nanomaterials for use in nanotechnology.[20] Nanobiotechnology can change the permeation of food packaging, improve mechanical and heat resistance, improve barrier properties, creating nano-biodegradable packaging materials as well as active antimicrobial surfaces. The discovery of biosensors has so many benefits that could generate unique products; by functional encapsulation, nanomaterials have the prospect of transforming the food industries. This has been employed in the development of novel biosensors for detecting pathogenic and toxic contamination. This new technique also generates significant concerns about the toxicity implications of nanotechnology in food, particularly for risk evaluation and safety. This also illustrates the desire for a new policy system capable of handling possible dangers related to nanoparticle deployment in food technology.[21]

Nanotechnology applications in different areas of the food industry have also developed as the research into using nanomaterials in food research has progressed. The food sector can use nanotechnology to improve food processing, production, quality control and storage. Despite traditional microscale components, nanoparticles with innovative properties might enhance food sensory acceptability by contributing new colours, appearances and textures. Nanotechnology has been employed to develop nanosensors for detecting dangerous substances in meals as well as to provide an intelligent packaging scheme that can detect the contamination of food efficiently and accurately. Nanoencapsulation is perhaps the most important food research innovation, particularly for bioactive chemicals and taste. After oral administration, controlled delivery devices using nanoencapsulation may improve the accessibility of bioactive substances. Furthermore, nanoencapsulation allows for the controlled release of aromas at the time specified, as well as the prevention of loss of taste between storage and processing. Recent uses of nanotechnology in food technology, such as flavour management and increased bioavailability of bioactive components, including monitoring of harmful elements in foods, have been highlighted in this study. In addition, this chapter addresses the categorization of nanocomposites, preparation techniques, and safety concerns in the area of food technology.[22]

Food safety concerns are a serious challenge in several industries including the agricultural, dairy, food and other industries that deal with foods due to their importance to human health. Current food safety problems (such as the melamine scandal in 2007–2008), as well as impacts on human health regarding chemically synthesized dietary supplements and chemical contaminants in food, have also prompted the development of sensitive, rapid and dependable technologies to identify certain food dangers. The continued advancement of nanotechnology, which could have benefits in detecting food components easily and quickly, provides an option. By combining nanotechnology with modern communication and information technologies, innovative and accessible methods for detecting various materials with good efficacy can be developed. To address the numerous criteria in food inspection, many kinds of nanosensors are already being created (nanosensors for identifying internal and external conditions in food packaging).[23]

The current global population is about 6 billion people; as a result of this massive population increase, food must be produced in a much safer, efficient, and sustainable manner, and it must be free of the harmful effects of infectious microbes. A considerable section of the population in advanced nations faces frequent shortages of food as nothing more than a response to environmental damage or other factors such as political instability, although food is plentiful in advanced countries. The goal of advanced countries should be to generate pest- and drought-resistant foods with high yields. The food sector in industrialized countries is reliant on consumer spending for fresher, cleaner foodstuffs; the utilization of nanotechnology in food technology is therefore discussed in this chapter.[24]

2.3 Relevance of Biological Sensors in the Food Industry

The surveillance of food quality and safety is a major global concern. It is vital to monitor food safety and quality to ensure that products that reach consumers are safe. The conventional analysis approaches for food standards and quality control, such as high-performance liquid chromatography and gas chromatography, are currently widely accepted and considered the best standards, however they are time-consuming, require expensive instrumentation and are inefficient. Biosensors, on the other hand, have become useful tools in agro-food diagnostics because they are portable, require less skilled operators, are convenient, and require no special expertise to operate. Low-quality food has been linked to a high rate of illness, morbidity, mortality and economic burden, and has become a major global public health concern. Electrochemical measurements are feasible choices for selective analyte measurement as required in the food industry. The use of biosensors is a novel approach to solving various challenges related to food packaging and various environmental issues.[25]

It is worth noting that biosensor sensitivity is not restricted to the detection of a single element. Many biosensors use biochemical pathways such as acetylcholinesterase inhibitors, which are employed to detect various pesticide contaminants even at minute levels in foods and food products. This indicates that biosensors can be used to detect an entire group of pesticides. Biosensors also have been developed to detect particular substances and toxin vectors that can cause acute toxicity. Checking for these pollutants is vital for food standards since they can have severe consequences even in low quantities.[26,27] Nanotechnology is an alternative means of improving food quality and flavour, by improving taste delivery and food consistency.[28] It is employed in the food industry primarily as nanosensors and nanostructured components to enhance production and packaging. Nanostructures include antimicrobial agents, additives, anticaking agents, and nutrient delivery and food packaging materials.[29] During food preparation, nanotechnology has been found to have various impacts by improving the shelf-life, texture, nutritional value, flavour and appearance of food. Surprisingly, nanotechnology also affects all of the aforementioned characteristics, but it has also resulted in substantial changes in food products, giving them remarkable characteristics.[15]

2.4 Nanosensors for Food Preservation

Strawberries have a short shelf-life after harvest, with their rate of metabolic activity and susceptibility to mould attacks especially (*Botrytis cinerea*) reducing fruit quality. Fakhouri *et al.*[30] investigated the efficacy of using cellulose nanocrystals that increased the shelf-life of strawberry fruit by 8 days, in which 5 g of gelatine was mixed with 100 mL of distilled water and various concentrations of cellulose nanocrystal (CNC) dispersions were added for 1 hour to yield the filmogenic solution; the results showed a significantly longer shelf-life with the treatment. For example, after eight storage sequences, the weight loss ranged from 31 percent to 36 percent in the covered samples. The edible coating showed success in retaining ascorbic acid (AA) in strawberries; while the AA content in control samples decayed rapidly, the AA concentration in the covered samples decayed slowly. The gel/CNC edible covering also had an antibacterial impact on the fruits.

Shi *et al.*[31] described an *in situ* sol–gel procedure that produced a unique chitosan nanosilica hybrid film on *Dimocarpus longan* fruit preserve using electron microscopy at room temperature. It was observed that there was an increase in the activity of polyphenol oxidase (PPO) and the activity of malondialdehyde of the improved fresh *Dimocarpus longan* compared to the unimproved one; they also observed an increase in its shelf-life along with delayed weight loss. Another viable alternative for improving the quality of fresh *Dimocarpus longan* fruit for long-term storage using a chitosan nanosilica coating has been illustrated.

Yu *et al.*[32] evaluated the effects of a 1 percent chitosan coating containing 0.04 percent nanosilicon dioxide on the qualitative attributes of harvested jujube at room temperature. In comparison to the control, the coated jujubes had lower red index, decay incidence, weight loss and respiration rate after
32 days. The chemical coating was responsible for the coated jujubes' decreased activity from phenylalanine ammonia-lyase and enhanced activity of scavenging antioxidant enzymes (superoxide peroxidase, catalase and dismutase). If nanosilicon dioxide is approved for food consumption, coating jujubes with chitosan plus nanosilicon dioxide could be a viable alternative.

The preservation of Coalho cheese with a nano-laminate covering was carried out using a sheet process. The Coalho cheese was tested for its microbial count, lipid peroxidation, mass loss, pH and titratable acidity for 20 days. The gas barrier and antibacterial capabilities of alginate and lysozyme nanocoating were discovered to have assisted the extension of the Coalho cheese shelf-life.[33] A high-energy emulsification–evaporation approach that relies on a 23-level factorial design was used to create nanoemulsions of β-carotene. According to the findings of Kong *et al.*,[34] it is conceivable to obtain suspensions at the nanoscale. During the 21-day storage period, the nanoemulsions maintained good stability. Preservation of size distribution was used to assess stability. Storage, on the other hand, affected β-carotene retention and colour in the micelles. The storage stability was affected by the processing conditions also. The presence of oxygen can result in the oxidation of antioxidants, which promotes bacterial growth. Methylene blue or titanium dioxide hybrid nanocomposite material has been used to produce colorimetric oxygen predictors for this purpose.[35] In most cases, trimethylamine is formed as a result of animal protein metabolism over time. As a result, the occurrence of trimethylamine is a sign of dehydration.[36]

2.5 Exposure to Food-pathogenic Bacteria and Food-contaminating Toxins

The genetic information of bacteria is the most common method for detecting pathogenic bacteria in food; nanoparticles can be used to help isolate the bacteria deoxyribonucleic acid (DNA) and detect bacteria in lower numbers and more effectively than existing methods. The DNA of a milk pathogenic bacteria (*Listeria monocytogenes*) was isolated using magnetic iron oxide nanoparticles. Conventional, droplet digital, touchdown and real-time quantitative PCR-based methods have been used for the molecular identification of microorganisms such as *Listeria monocytogenes* and *Listeria ivanovii* from food and environmental sources.[37] Molecular identification of these microbes using 16 s ribosomal ribonucleic acid is widely utilized as a selective marker. This method (microarray approach based) lacks sensitivity and is quite expensive. A large amount of DNA sequences can be studied at the same time when using microarrays or microbiomes, allowing the enhancement of characterization and identification of many pathogenic microbial populations using their genes in a single assay.[38] Rapid and very sensitive detection devices have resulted from thin film-based optical immunosensors that generate signals for the detection of microbial cells.[39] Common foodborne pathogens such as *E. coli* (O157:H7) and *Staphylococcus aureus* were detected using a dimethylsiloxane microfluidic immunosensor and electrochemical impedance spectrum with particular antibodies mounted on an alumina nanoporous membrane.[40] Aflatoxins are a class of poisonous and carcinogenic chemicals discovered in *Aspergillus flavus*- and *Aspergillus parasiticus*-contaminated foods.

Clostridium botulinum and related *Clostridia* spp could produce neurotoxins that could induce severe, potentially fatal poisoning in humans which are for detection and characterization of these neurotoxins. Vanhomwegen *et al.*[41] used a resequencing microarray and the pathogen in combination with an automated data method. This method enabled the identification of represented strains of various subtypes and toxin types in a naturally contaminated food sample, as well as the neurotoxin-producing *C. botulinum* variant. Furthermore, this approach enabled detailed analysis of the various neurotoxin gene cluster components in all strains tested. Aflatoxins B1 and M1 have been detected with enzyme-linked and immunosorbent immunoassays.[42] Electrochemical methods such as cyclic voltammetry and impedance have been used to detect Ochratoxin-A, cerium dioxide, gold–graphene oxide–ionic liquid and zinc oxide.[43]

2.6 Nanobarcodes for Product Authenticity

Product authentication labels employ barcodes all over the world, however due to their two-dimensional barcodes they are more vulnerable to damage, ease of location manipulation and falsification. On the other hand, invisible barcodes made of nanoparticles are difficult to alter.[44] Recent research has shown a novel encoding technology using disc codes (nanodiscs), which is a nanoparticle-based technology. The use of nanodisc protocols involves the application of Raman microscopic equipment for scanning a series of surface-enhanced Raman scattering disc pairs.[45] To analyse, anthrax, tularaemia, *E. coli* and some severe acute respiratory viral infections, fluorescence microscopy, flow cytometry and dot blotting have been studied where the analyte was a barcode and DNA, and the technique used was scanning confocal Raman spectroscopy.[46]

2.7 Electronic Nose and Tongue

An electronic nose is a gadget that uses aroma-detecting techniques to detect scents. It has an olfactory system similar to that of mammals. The electronic nose is mostly made up of gas sensors that can detect changes in odour and flavour, and also quantity and quality. A larger surface area of nanoparticles than macroscopic particles aids in the greater gas absorption on sensor surfaces.[47,48] The formation of a carboxylated polypyrrole (C-Ppy) nanocomposite field-effect device with hTAS2R38, a specific human receptor for a bitter taste, led to the development of a nanobioelectronic sensor used in sensing bitter tastes. Interestingly, the nanotongue was able to detect propylthiouracil and phenylthiocarbamide bitterness chemicals, having a capacity close to human sensitivity.[49]

Gas detection is important in vegetable and fruit monitoring during processing and storage. The quality of these foodstuffs degrades when they are exposed to excessive ethylene gas. For ethylene sensing, tungsten oxide–tin oxide nanoparticles have been used.[50] A collection of quartz crystal microbalance detectors was augmented with gold nanoparticles to create an electronic nose. Non-edible lampante olive oil and extra virgin oil were detected also using this specially constructed nose.[51] In the same way, nanocrystals were employed to assess the quality of 17 common Chinese vinegars.[52] The bioelectronic tongue was able to distinguish bitter from non-bitter taste at molar concentrations.[53] A bioelectronic tongue was developed using a single-walled carbon nanotube with a surface modified using G-protein, making it possible to distinguish between bitter and sweet taste.[54]

2.8 Safety Issues

Besides its many great uses, concerns over the safety of food using nanotechnology still require attention. Several researchers have explored nanostructured materials safety concerns, discussing the health influence they may have on consumers when nanoparticles migrate from the packaging into the food.[55,56] Risk assessments for all nanosystems must be evaluated, even if the substance is recognized as safe because the physiochemical state of nanosystems differs from its macro-states as there could be a danger of bioaccumulation of these nanomaterials in organs and tissues.[57] If human cells are exposed to nanofillers such as anti-caking agents, it could lead to cytotoxicity,[58] or microplastics used may result in detrimental health outcomes.[59] The morphology of particles, concentration, adsorption aggregation and surface energy are some of the elements which influence dissolution.[60] Silver–copper nanocomposite diffusion confirms the proportion of nanofiller present in these nanocomposites as well as particle size, contact, temperature and duration, which are among the most important factors driving migration. Each nanomaterial has its own unique properties and toxicity which help to determine the level of risk. To conclude, the popularity of nanoscale structures in the food industry has grown over the years, and has led to significant interest in this area of study. Advances in nano-based biotechnology have resulted in the utilization of smaller and more sensitive devices, which allows them to be used for improvements of the appearance, texture and taste of food. They are also used for the detection of aroma, bitter taste, pathogens and toxins in the food industry.[19]

Nanoparticle-based biosensors in the food sector present a very quick, reliable, attractive and easy method for monitoring or analysing the quality and safety of products. These biosensors have transformed the food industry by reducing fears about food safety. However, the use of these devices and gadgets still raises concerns despite their enormous advantages. There are still concerns raised regarding the biosafety of the nanomaterials used for many of these biosensors, with continuous research on-going to provide more environmental nanoparticle-based biosensors that are biocompatible and have very low toxicity.

References

1. M. Lamri, T. Bhattacharya, F. Boukid and M. Gagaoua, Nanotechnology as a processing and packaging tool to improve meat quality and safety, *Foods*, 2021, **10**(11), 2633.
2. C. O. Adetunji, S. A. Akande, A. K. Oladipo, R. A. Salawu and A. F. Onyegbula, Determination of the microbiological quality and proximate composition of fermented cassava food products sold in ilorin-west local government area, Nigeria, *Ruhuna j. Sci.*, 2017, **8**, 76–89.
3. F. F. Olayemi and C. O. Adetunji, Effect of rinses on microbial quality of commercially available eggs and their components before processing from Ilorin in western Nigeria, *Bitlis Eren Univ. J. Sci. Technol.*, 2013, **3**(2), 44–47.
4. C. O. Adetunji, O. B. Fawole, K. A. Arowora, S. I. Nwaubani, E. S. Ajayi, J. K. Oloke, J. A. Aina, J. B. Adetunji and A. O. Ajani, Postharvest quality and safety maintenance of the physical properties of daucus *carota L.* fruits by neem oil and moringa oil treatment: A new edible coatings, *Agrosearch*, 2013, **13**(1), 131–141.
5. C. O. Adetunji, S. M. Olugbenga, K. Oseni, V. Ajit, A. Muhammad, K. O. Julius, S. Hamda, B. A. Juliana, J. Aditi, E. B. Ruth, O. Phebean and E. U. Benjamin, Quinoa: From Farm to Traditional Healing, Food Application, and Phytopharmacology, in *Biology and Biotechnology of Quinoa*, ed. A. Varma, Springer, Singapore, 2021.
6. C. O. Adetunji, S. M. Olugbenga, N. Wilson, E. U. Kingsley, A. A. Osikemekha, K. O. Julius, V. Ajit, K. Oseni, J. Aditi and B. A. Q. Juliana, The next biotech plant: Food security and environmental and health hot spots, in *Biology and Biotechnology of Quinoa*, ed. A. Varma, Springer, Singapore, 2021.
7. C. O. Adetunji, S. M. Olugbenga, V. Ajit, K. O. Julius, K. Oseni, A. Muhammad, E. B. Ruth, I. Areeba, B. A. Juliana, S. Khuram, J. Aditi, E. U. Benjamin, M. Noshiza, O. Phebean and N. O. Frances, Recent advances in the application of biotechnology for improving the production of secondary metabolites from quinoa, in *Biology and Biotechnology of Quinoa*, ed. A. Varma, Springer, Singapore, 2021.

8. C. O. Adetunji, T. O. Olugbemi, O. Adeyomoye, A. Dare, M. J. Adeniyi, E. Alex, M. Rebezov, L. Garipova and M. A. Shariati, *et al.*, eHealth, mHealth, and telemedicine for COVID-19 pandemic, in *Assessing COVID-19 and Other Pandemics and Epidemics using Computational Modelling and Data Analysis*, ed. S. K. Pani, S. Dash, W. P. dos Santos, S. A. Chan Bukhari and F. Flammini, Springer, Cham, 2022.
9. O. A. Charles, T. O. Olugbemi, O. Adeyomoye, A. Dare, M. J. Adeniyi, E. Alex, M. Rebezov, E. Petukhova and A. S. Mohammad, Machine learning approaches for COVID-19 pandemic, in *Assessing COVID-19 and Other Pandemics and Epidemics using Computational Modelling and Data Analysis*, ed. S. K. Pani, S. Dash, W. P. dos Santos, S. A. Chan Bukhari and F. Flammini, Springer, Cham, 2022.
10. O. A. Charles, T. O. Olugbemi, O. Adeyomoye, A. Dare, M. J. Adeniyi, E. Alex, M. Rebezov, O. Isabekova and A. S. Mohammad, Smart Sensing for COVID-19 Pandemic, in *Assessing COVID-19 and Other Pandemics and Epidemics using Computational Modelling and Data Analysis*, ed. S. K. Pani, S. Dash, W. P. dos Santos, S. A. Chan Bukhari and F. Flammini, Springer, Cham. 2022, pp. 145–156.
11. O. A. Charles, T. O. Olugbemi, O. Adeyomoye, A. Dare, M. J. Adeniyi, E. Alex, M. Rebezov, E. Petukhova and A. S. Mohammad, Internet of Health Things (IoHT) for COVID-19, in *Assessing COVID-19 and Other Pandemics and Epidemics using Computational Modelling and Data Analysis*, ed. S. K. Pani, S. Dash, W. P. dos Santos, S. A. Chan Bukhari and F. Flammini, Springer, Cham, 2022, pp. 75–87.
12. O. A. Charles, T. O. Olugbemi, A. Olorunsola, A. Dare, M. J. Adeniyi, E. Alex, M. Rebezov, K. Natalia and A. S. Mohammad, Diverse techniques applied for effective diagnosis of COVID-19, in *Assessing COVID-19 and Other Pandemics and Epidemics using Computational Modelling and Data Analysis*, ed. S. K. Pani, S. Dash, W. P. dos Santos, S. A. Chan Bukhari and F. Flammini, Springer, Cham, 2022, pp. 45–58.
13. J. Jeevanandam, A. Barhoum, Y. Chan, A. Dufresne and M. Danquah, Review on nanoparticles and nanostructured materials: history, sources, toxicity and regulations, *Beilstein J. Nanotechnol.*, 2018, **9**, 1050–1074.
14. S. H. Nile, V. Baskar, D. Selvaraj, A. Nile, J. Xiao and G. Kai, Nanotechnologies in food science: Applications, recent trends, and future perspectives, *Nano-Micro Lett.*, 2020, **12**, 45.
15. T. Sing, S. Shukla, P. Kumar, V. wahla, K. Vivek and I. Rather, Application of nanotechnology in food science: Perception and overview, *Front. Microbiol.*, 2017, **8**, 1501.
16. T. King, M. Osmond and L. Duffy, Nanotechnology in the food sector and potential applications for the poultry industry, *Trends Food Sci. Technol.*, 2018, **72**, 62–73.
17. S. Mehwish, A. Sumaira, H. Christophe, A. Iram and H. A. Bilal, An overview of the applications of nanomaterials and nanodevices in the food industry, *Food*, 2020, **9**(2), 148.

18. D. Renata, Metal nanoparticles in nanosensors for food quality assurance, *Scientific J. Logistics*, 2020, **16**(2), 271–278.
19. V. Kumar, P. Guleria and S. K. Mehta, Nanosensors for food quality and safety assessment, *Environ. Chem. Lett.*, 2017, **15**, 165–177.
20. T. Caon, Nanobiosensors: New trends in the food industry: Application of nanosensors in food packaging, *Nanobiosensor*, 2017, **22**(7), 773–804.
21. M. M. Berekaa, Nanotechnology in food industry; advances in food processing, packaging and food safety, *Int. J. Curr. Microbiol. App. Sci.*, 2015, **4**(5), 345–357.
22. Y. Hyunjong, P. Jun-Young, W. K. Chang, H. Sung-Chul, P. Kyung-Min and C. Pahn-Shick, An overview of nanotechnology in food science: Preparative methods, practical applications, and safety, *J. Chem.*, 2018, **22**(4), DOI: 10.1155/2018/5427978.
23. O. M. Enisa and M. Mirjana, Nanosensors applications in agriculture and food industry, *Bull. Chem. Technol. Soc. Bosnia Herzegovina*, 2016, **47**, 59–70.
24. S. K. Ameta, A. K. Rai, D. Hiran, R. Ameta and S. C. Ameta, Use of nanomaterials in food science, *Biog. Nanopart. Agroecosystems*, 2020, 457–488.
25. J. Bunney, S. Williamson, D. Atkin, M. Jeanneret, D. Cozzolino and J. Chapman, The use of electrochemical biosensors in food analysis, *Curr. Res. Nutr. Food Sci. J.*, 2017, **5**(3), 183–195.
26. H. Turasan and J. Kokini, Novel nondestructive biosensors for the food industry, *Annu. Rev. Food Sci. Technol.*, 2021, **12**(1), 539–566.
27. T. M. Oliveira, M. F. Barroso, S. Morais, P. de Lima-Neto, A. N. Correia, M. B. Oliveira and C. Delerue-Matos, Biosensor based on multi-walled carbon nanotubes paste electrode modified with laccase for pirimicarb pesticide quantification, *Talanta*, 2013, **106**, 137–143.
28. K. Nakagawa, Nano- and micro-encapsulation of flavor in food systems, in *Nano- and Microencapsulation for Foods*, ed. H.-S. Kwak, John Wiley & Sons, Oxford, 2014, ch. 10, pp. 249–272.
29. P. N. Ezhilarasi, P. Karthik, N. Chhanwal and C. Anandharamakrishnan, Nanoencapsulation techniques for food bioactive components: a review, *Food Bioprocess Technol.*, 2013, **6**, 628–647.
30. F. M. Fakhouri, A. C. A. Casari, M. Mariano, F. M. Fakhouri, A. C. A. Casari, M. Mariano, F. Yamashita, L. I. Mei, V. Soldi and S. M. Martelli, Effect of a gelatin-based edible coating containing cellulose nanocrystals (CNC) on the quality and nutrient retention of fresh strawberries during storage, in *Proceedings of the IOP Conference Series: Materials Science and Engineering, Conference 1, 2nd International Conference on Structural Nano Composites (NANOSTRUC 2014)*, Madrid, 2014, vol. 64, DOI: 10.1088/1757-899X/64/1/012024.
31. S. Shi, W. Wang, L. Liu, S. Wu, Y. Wei and W. Li, Effect of chitosan/nano-silica coating on the physicochemical characteristics of longan fruit under ambient temperature, *J. Food Eng.*, 2013, **118**, 125–131.

32. Y. Yu, S. Zhang, Y. Ren, H. Li, X. Zhang and J. Di, Jujube preservation using chitosan film with nano-silicon dioxide, *J. Food Eng.*, 2012, **113**, 408–414.
33. B. G. D. S. Medeiros, M. P. Souza, A. C. Pinheiro, A. I. Bourbon, M. A. Cerqueira and A. A. Vicente, *et al.*, Physical characterisation of an alginate/lysozyme nano-laminate coating and its evaluation on 'Coalho' cheese shelf life, *Food Bioproc. Technol.*, 2014, 7, 1088–1098.
34. M. Kong, X. G. Chen, D. K. Kweon and H. J. Park, Investigations on skin permeation of hyaluronic acid based nanoemulsion as transdermal carrier, *Carbohydr. Polym.*, 2011, **86**, 837–843.
35. D. Gutierrez-Tauste, X. Domenech, N. Casan and J. Ayllon, Characterization of methylene blue/TiO_2 hybrid thin films prepared by the liquid phase deposition (LPD) method: application for fabrication of light-activated colorimetric oxygen indicators, *J. Photochem. Photobiol., A*, 2007, **187**, 45–52.
36. W. H. Zhang and W. D. Zhang, Fabrication of SnO_2–ZnO nanocomposite sensor for selective sensing of trimethylamine and the freshness of fishes, *Sens. Actuators, B*, 2008, **134**, 403–408.
37. H. Yang, L. Qu, A. Wimbrow, X. Jiang and Y. Sun, Rapid detection of *Listeria monocytogenes* by nanoparticle-based immunomagnetic separation and real-time PCR, *Int. J. Food Microbiol.*, 2007, **118**, 132–138.
38. A. Rasooly and K. Herold, Food microbial pathogen detection and analysis using DNA microarray technologies, *Foodborne Pathog. Dis.*, 2008, **5**(4), 531–550.
39. A. Subramanian, A mixed self-assembled monolayer-based surface Plasmon immunosensor for detection of E. coli O157H7, *Biosens. Bioelectron.*, 2006, 7, 998–1006.
40. F. Tan, P. H. Leung, Z. B. Liu, Y. Zhang, L. Xiao, W. Ye, X. Zhang, L. Yi and M. A. Yang, PDMS microfluidic impedance immunosensor for E. coli O157: H7 and Staphylococcus aureus detection via antibody-immobilized nanoporous membrane, *Sens. Actuators, B*, 2011 Nov 28, **159**(1), 328–335.
41. J. Vanhomwegen, N. Berthet, C. Mazuet, G. Guigon, T. Vallaeys, R. Stamboliyska, P. Dubois, G. C. Kennedy, S. T. Cole, V. Caro and J. C. Manuguerra, Application of high-density DNA resequencing microarray for detection and characterization of botulinum neurotoxin-producing clostridia, *PLoS One*, 2013, **8**(6), e67510.
42. C. Zhang, A. X. Yin, R. Jiang, J. Rong, L. Dong, T. Zhao, L. D. Sun, J. Wang, X. Chen and C. H. Yan, Time temperature indicator for perishable products based on kinetically programmable Ag overgrowth on Au nanorods, *ACS Nano*, 2013, 7, 4561–4568.
43. P. Norouzi, B. Larijani and M. R. Ganjali, Ochratoxin a sensor based on nanocomposite hybrid film of ionic liquid-graphene nano-sheets using coulometric FFT cyclic voltammetry, *Int. J. Electrochem. Sci.*, 2012, 7, 7313–7324.
44. Y. C. Wang, L. Lu and S. Gunasekaran, Gold nanoparticle-based thermal history indicator for monitoring low-temperature storage, *Microchim. Acta*, 2015, **182**, 1305–1311.

45. L. Qin, M. J. Banholzer, J. E. Millstone and C. A. Mirkin, Nanodisk codes, *Nano Lett.*, 2007, 7, 3849–3853.
46. D. Liu, W. Chen, J. Wei, X. Li, Z. Wang and X. Jiang, A highly sensitive, dual-readout assay based on gold nanoparticles for organophosphorus and carbamate pesticides, *Anal. Chem.*, 2012, **84**, 4185–4191.
47. S. Ranjan, N. Dasgupta, A. R. Chakraborty, S. M. Samuel, C. Ramalingam, R. Shanker and A. Kumar, Nanoscience and nanotechnologies in food industries: opportunities and research trends, *J. Nanopart. Res.*, 2014, **16**, 2464.
48. N. Dasgupta, S. Ranjan, D. Mundekkad, C. Ramalingam, R. Shanker and A. Kumar, Nanotechnology in agro-food: From field to plate, *Food Res. Int.*, 2015, **69**, 381–400.
49. H. S. Song, O. S. Kwon, S. H. Lee, S. J. Park, U. K. Kim, J. Jang and T. H. Park, Human taste receptor-functionalized field effect transistor as a human-like nanobioelectronic tongue, *Nano Lett.*, 2013, **13**, 172–178.
50. Y. Pimtong-Ngam, S. Jiemsirilers and S. Supothina, Preparation of tungsten oxide–tin oxide nanocomposites and their ethylene sensing characteristics, *Sens. Actuators, A*, 2007, **139**, 7–11.
51. M. D. Carlo, G. C. Fusella, A. Pepe, M. Sergi, M. D. Martino, M. Mascini, G. Martino, A. Cichelli, C. D. Natale and D. Compagnone, Novel oligopeptides-based e-nose for food quality control: application to extra-virgin olive samples, *Qual. Assur. Saf. Crops Foods*, 2014, **6**, 307–317.
52. T. Zhang, C. Lv, L. Chen, G. Bai, G. Zhao and C. Xu, Encapsulation of anthocyanin molecules within a ferritin nanocage increases their stability and cell uptake efficiency, *Food Res. Int.*, 2014, **62**, 183–192.
53. T. H. Kim, H. S. Song, H. J. Jin, S. H. Lee, S. Namgung, U. K. Kim, T. H. Park and S. Hong, Bioelectronic super-taster device based on taste receptor-carbon nanotube hybrid structures, *Lab Chip*, 2011, **11**, 2262–2267.
54. H. S. Song, H. J. Jin, S. R. Ahn, D. Kim, S. H. Lee, U. K. Kim, C. T. Simons, S. Hong and T. H. Park, Bioelectronic tongue using heterodimeric human taste receptor for the discrimination of sweeteners with human-like performance, *ACS Nano*, 2014, **8**, 9781–9789.
55. E. L. Bradley, L. Castle and Q. Chaudhry, Applications of nanomaterials in food packaging with a consideration of opportunities for developing countries, *Trends Food Sci. Technol.*, 2011, **22**(11), 604–610.
56. A. Jain, R. Shivendu, D. Nandita and R. Chidambaram, Nanomaterials in food and agriculture: an overview on their safety concerns and regulatory issues. Sensor, *Sens. Actuators, B*, 2016, **159**, 328–335.
57. K. Savolainen, L. Pylkkänen, H. Norppa, G. Falck, H. Lindberg, T. Tuomi, M. Vippola, H. Alenius, K. Hämeri, J. Koivisto and D. Brouwer, Nanotechnologies, engineered nanomaterials and occupational health and safety–A review, *Saf. sci.*, 2010 Oct 1, **48**(8), 957–963.
58. J. Athinarayanan, V. S. Periasamy, M. A. Alsaif, A. A. Al-Warthan and A. A. Alshatwi, Presence of nano silica (E551) in commercial food products: TNF-mediated oxidative stress and altered cell cycle

progression in human lung fibroblast cells, *Cell Biol. Toxicol.*, 2014, **30**, 89–100.

59. C. O. Adetunji, A. Inobeme, J. Tadso, O. T. Olaniyan, O. F. Abimbola, M. Shahnawaz and O. Anani, Potential of plastic waste in enhancing the level of pathogenicity of diverse pathogens in the marine biota, *Impact of Plastic Waste on the Marine Biota*, Springer, Singapore, 2022, pp. 301–312.
60. M. Cushen, J. Kerry, M. Morris, M. Cruz-Romero and E. Cummins, Evaluasian and simulation of silver and copper nanoparticle migration from polyethylene nanocomposites to food and an associated exposure assessment, *J. Agric. Food Chem.*, 2014, **62**, 1403–1411.

CHAPTER 3

Bio-based Sensing: Role of Natural Dyes in Food Freshness Indicators

V. G. MARTINS,*[a] L. G. SANTOS,[a] V. P. ROMANI[a,b] AND S. S. FERNANDES[a,c]

[a] Laboratory of Food Technology, School of Chemistry and Food Engineering, Federal University of Rio Grande, Rio Grande – RS 96203-900, Brazil; [b] Department of Food Engineering, State University of Midwest, Guarapuava – PR 85040-167, Brazil; [c] Center of Social Sciences, Health and Technology, Federal University of Maranhão, Imperatriz – MA 65900-410, Brazil
*Email: vilasiamartins@gmail.com

3.1 Introduction

Dyes are a class of food additives, defined as substances that are intended to confer, intensify or standardize the coloring of food products. There are natural dyes that are obtained from vegetables, animals or microorganisms; artificial dyes, which are those obtained through a synthetic process, with a defined chemical composition; and synthetic dyes identical to natural dyes, which are those whose chemical structures are similar to natural dyes, but that are synthesized in the laboratory.

Artificial dyes are a class of chemical additives with no nutritional value. From a toxicological point of view, several studies have been carried out to verify their harmful effects on humans, as these additives are not completely

Food Chemistry, Function and Analysis
Bio- and Nano-sensing Technologies for Food Processing and Packaging
Edited by Ashutosh Kumar Shukla

Published by the Royal Society of Chemistry, www.rsc.org

harmless to health. The use of artificial dyes is advantageous mainly due to the greater dyeing power, stability, uniformity and lower cost, when compared to natural dyes. However, despite all these positive points, artificial dyes have been identified as harmful to health, being associated with the emergence of some diseases.

In view of the above, the trend is to reduce the use of artificial dyes. The natural dye industry is investing in research to improve light and heat stability and extraction cost which still limit their use. Despite these limitations, natural dyes have been a trend in the food industry, as they give a natural aspect to products, gaining greater acceptance by consumers. Natural dyes come in a wide variety of shades and when mixed can form a huge color palette as large as that of artificial ones. Among them, the most common are carotenoids, betalains, chlorophylls, anthocyanins, curcumins and phycocyanins. They are present in various types of products, such as fruits and vegetables, algae, roots, leaves, among others, making them easily obtainable. These dyes can be used without any health risk in a variety of food products and as indicators of intelligent packaging.

Intelligent packaging informs consumers about the quality of the food product they are storing, in this case a colorimetric sensor/indicator changes color when it is spoiled, and the product then must not be consumed. The natural dyes widely used in the development of biodegradable freshness indicators are chlorophylls, anthocyanins, betalains, curcumins, quercetins and phycocyanins, which can be applied in a single form or a mixture of these.[42,46,50] The packages developed by the application of natural dyes have great colorimetric sensitivity to changes in the quality of the packaged product, such as change in pH, release of ammonia gas and also exposure to inadequate storage temperature and UV-light.[50]

In the literature, studies have explored the incorporation of these natural dyes in various polymeric matrices, including synthetic (*e.g.* polyvinyl alcohol, polyethylene oxide, polyethylene, *etc.*) and natural polymers (*e.g.* starch, cellulose, chitosan, pectin, *etc.*). Biopolymers produced with polysaccharides, proteins and lipids incorporated with natural dyes to generate intelligent films contribute, besides the indication of spoilage or freshness of the products, to the decrease in the use of conventional synthetic plastics, which are causing environmental problems. The application of intelligent films produced with natural dyes has been explored mostly in meat products, fish and milk.

This chapter illustrates the variety of natural dyes that can be used in sensors/indicators for intelligent packaging, their color variations in different media, potential applications and the current research being carried out in this field of food packaging.

3.2 Natural Dyes

Currently, the increase in environmental threats and health concerns has made many researchers focus their research into alternatives to control pollution using sustainable approaches.[1,2] Associated with this, industrial food production is one of the most prevalent activities worldwide and requires constant innovations in product development.[3]

Over the last few decades, natural dyes have been used to replace synthetic dyes in food, reducing the harmful effects caused by allergy and intolerance of synthetic pigments.[4] Associated with this, the use of synthetic dyes in food packaging harms the environment and human health. To solve this problem studies have been focusing on the use of natural dyes[5] and a recent potential application has been the use of natural dyes as food freshness sensors.[6]

As they are non-toxic and non-carcinogenic, natural dyes are considered safe. In addition, some natural dyes still have anti-UV (ultraviolet light) and antimicrobial properties.[2] In this sense, natural dyes, which are ecological, sustainable and biodegradable, are increasingly gaining global popularity.[1,2] These compounds are readily available in nature and can be obtained from plants, animals and microorganisms.[7] The highest content of natural colorings is present in the outer layers, such as the skin, leaves and peel of fruits and vegetables. These parts are often described as food waste and generally form the base of the bagasse, being a low-cost source for obtaining natural dye and/or bioactive compounds.[2]

Several researchers have focused on the extraction and application of natural dyes from different sources. Table 3.1 presents the main natural dyes that have been recently studied.

The use of dyes in the food industry represents 36% of the global dyestuffs market and its use aims to confer or enhance the color of food products, making them more attractive to the consumer.[22] The natural dyes that stand out are those that include chlorophylls, carotenoids, phycocyanins, betalains, anthocyanins and other phenolic compounds.[23] Among these,

Table 3.1 Main natural dyes used in the food sector.

Natural dyes	Compounds	Colors	References
Betalains	Betacyanin	Red	Qin *et al.* (2020)[6]
	Betaxantina	Yellow	Mayolo-Deloisa *et al.* (2020)[8]
	β-Carotene	Orange-yellow	Stupar *et al.* (2021)[9]
	Lutein	Orange-yellow	Chen *et al.* (2021)[10]
Carotenoids	Lycopene	Yellow to red	Dhakane-Lad and Kar (2021)[11]
	Capsanthin	Red	Jiménez *et al.* (2021)[12]
	Astaxanthin	Yellow to red	Wang *et al.* (2021)[13]
	Canthaxanthin	Orange, red	Abuthahir *et al.* (2021)[14]
	Zeaxanthin	Orange, red	Bourdon *et al.* (2021)[15]
	Cyanidin	Red, blue, or violet	Petruk *et al.* (2017)[16]
Anthocyanins	Malvidin		
	Peonidin		Rajan *et al.* (2018)[17]
Flavin	Riboflavin	Yellow	Zhang *et al.* (2021)[18]
Curcuminoid	Curcumin	Yellow	Aliabbasi *et al.* (2021)[19]
	Turmeric	Yellow	Mousavi *et al.* (2021)[20]
Tetrapyrrole	Chlorophylls	Green	Ezike *et al.* (2021)[21]

anthocyanins, betalains and carotenoids are the most used in the food industry due to their bright colors and good water solubility.[24]

Anthocyanins are phenolic compounds that belong to one of the six categories of the flavonoid class, with these being among the most colorful group and responsible for the colors red and purple.[2] Most plant species produce anthocyanins, especially in red, blue or purple fruits, such as berries and red grapes.[24] In addition to the most common sources of anthocyanins, such as grapes[25] and jaboticaba,[26] several other sources of anthocyanins have been studied, such as purple rice,[27] black corn, chokeberry,[28] purple-fleshed sweet potatoes,[29] purple onion peel[30] and blueberry.[31]

Betalains are tyrosine-derived compounds, synthesized from a nitrogenous core structure, called betalamic acid, that give them a red or yellow color. Betalains can be subdivided into betacyanins and betaxanthins.[24] These natural dyes are related to the maintenance and regulation of oxidation–reduction processes due to their high antioxidant capacity.[32]

Betalains are rare in nature, exclusively present in families of the Caryophyllales in a mutually exclusive manner with anthocyanins, as well as in some genera of higher fungi.[24] Betalains can be obtained through ulluco (*Ullucus tuberosus*),[33] prickly pears (*Oxalis dilleniid*),[34] red beet (*Beta vulgaris*),[24] vegetable amaranth (*Amaranthus tricolor* L.)[35] and cactus berry fruits (*Myrtillocactus geometrizans*).[36]

Carotenoids belong to the class of tetraterpenes with linear molecules, which consist of eight isoprenoid units that have an inversion center. The most frequently found carotenoids are lycopene, present mainly in red-fleshed fruits and fruit vegetables, and β-carotene, which is more ubiquitous but at low levels.[2] Carotenoids are popularly found in tomatoes and carrots, but there are other sources such as annatto seeds,[37] pumpkin,[38] mandarin[39] and microalgae.[40]

Many natural dyes are extracted from biomass, which generates a significant environmental impact at the agricultural and extraction stages. The type and concentration of dye depends on the biomass source. Vegetables and fruits are primary sources of natural dyes, as they usually contain high concentrations of these dyes. However, extracting dyes from primary sources is an expensive process, besides the food safety issues and the risk of deforestation. Thus, a sustainable source of biomass can be obtained from food waste produced during the processing steps of the industries.[2]

The use of some natural dyes, whether for application in food or intelligent packaging, can present technological limitations, as they are generally sensitive to oxidation and high temperatures, which leads to a marked discoloration during processing and storage, reducing the sensory quality of the products and the action of these dyes. Thus, it becomes necessary to use techniques that increase their stability.[3]

Micro- and nanoencapsulation are processing techniques that can be applied to natural dyes to increase their stability, as they limit their interaction with air and other food constituents. This technique also minimizes the effects of high temperatures on dyes, reducing water activity (which reduces microbial growth and storage or transport costs), and allows the use of these compounds in foods where they would be incompatible due to pH or solubility conditions.[3]

3.3 Natural Dyes in Food Freshness Indicators

Environmental impacts caused by the increased use of synthetic materials have encouraged the study of the development of biodegradable materials with non-toxic properties and mechanical, barriers and transparency characteristics suitable to be applied as food packaging.[30] In this context, biodegradable packaging materials are functionalized with dyes or natural pigments, producing colorimetric indicators capable of monitoring the quality and freshness of packaged foods.[41,42]

Numerous studies have demonstrated the great ability that synthetic dyes (*e.g.*, bromothymol blue,[43] methyl red[44] and bromocresol purple[45]) have in the development of intelligent packaging. However, synthetic dyes can be potentially toxic and mutagenic compounds that, when incorporated into food packaging, their accidental release can contaminate the packaged product, making consumption unfeasible and increasing the generation of waste.[46] In this context, recent studies propose the development of smart or intelligent packaging from the incorporation of natural dyes, extracted from plants,[47] food waste[48] and cyanobacteria,[49] which indicate low levels of toxicity, economic viability, being derived from renewable sources and with easy access.

The natural dyes widely used in the development of biodegradable freshness indicators are chlorophylls, anthocyanins, betalains, curcumins, quercetins and phycocyanins, which can be applied in a single form or a mixture of these.[42,46,50] The packages developed by the application of natural dyes have great colorimetric sensitivity to changes in the quality of the packaged product, such as change in pH, release of ammonia gas and also exposure to inadequate storage temperature and UV light.[50] Table 3.2 shows studies on the application of natural dyes in the development of intelligent packaging as a colorimetric indicator of packaged food quality.

Colorimetric indicators functionalized with natural dyes have the ability to monitor food quality due to changes that can occur inside the package during storage.[41,46] Therefore, these indicators are able to react with changes in pH, gases (*e.g.* CO_2, O_2, HNO_3, ethylene, NH_3, *etc.*), pathogens and their metabolites and exposure to inadequate temperature and moisture of the product, indicating colorimetrically to the consumer aspects related to the quality and suitability for consumption of packaged food.

3.3.1 Anthocyanins

Anthocyanins are classified as phenolic compounds belonging to the class of flavanols and subclass of anthocyanidins.[65] Anthocyanins represent a large

Table 3.2 Natural dyes used in the development of intelligent packaging.

Natural dyes	Source	Stimulating factors	References
Anthocyanins	Bark, leaves, fruits, vegetables	pH, ammonia, and temperature	Filipini *et al.* (2020); Liang *et al.* (2018); Liu *et al.* (2019); MacIel *et al.* (2012); Pereira *et al.* (2015)[48,51–54]
Betalains	Bark, leaves, root, vegetables	pH and ammonia	Hu *et al.* (2020); Qin *et al.* (2020)[6,35]
Carotenoid	Bark, leaves, root, vegetables, microorganisms	Temperature and pH	Jamróz *et al.* (2019); Latos-Brozio and Masek, (2020)[55,56]
Chlorophyll	Bark, leaf, fruits, cyanobacteria	pH, ammonia, temperature, and oil oxidizing agents	Chavoshizadeh *et al.* (2020); Jancikova *et al.* (2019); Kuntzler *et al.* (2020)[49,57,58]
Curcumin	Root	pH and ammonia	Etxabide *et al.* (2021); Musso *et al.* (2017); Zhai *et al.* (2020)[59–61]
Phycocyanin	Cyanobacteria and red algae	pH	Moreira *et al.* (2018); Terra *et al.* (2021)[62,63]
Quercetin	Leaves, bark, fruits, vegetables	pH, temperature, and metals	Bai *et al.* (2019); Terra *et al.* (2021)[62,64]

group of water-soluble natural dyes found in many parts of plants.[66] This natural dye is responsible for the red, orange, blue and purple colors of vegetables, with great bioactive potential due to its antioxidant and antimicrobial properties.[46] The color stability of anthocyanins depends on several factors, such as storage temperature, enzymes, oxygen content and pH. The color change of anthocyanins caused by exposure to these factors is the result of changes that occur in the chemical structures of the pigment, enhancing its application in the development of intelligent packaging to assess the freshness and quality of food.[46,67]

More than 500 types of anthocyanins are known, the most commonly found being 3-*o*-glucosides or 3,5-*o*-diglucosides.[68] In general, anthocyanins are glycosylated to anthocyanidins, being basically composed of a six-membered heterocyclic ring containing an oxygen (C) between two aromatic rings (A–B) (see Scheme 3.1). Each type of anthocyanin differs according to the position of the hydroxyl and/or methoxy groups, glycosylated bond number and acetylation degree.[69] In addition, the color of each anthocyanin species can change depending on its source and chemical structure, larger amounts of methoxy groups cause a reddish coloration, while a

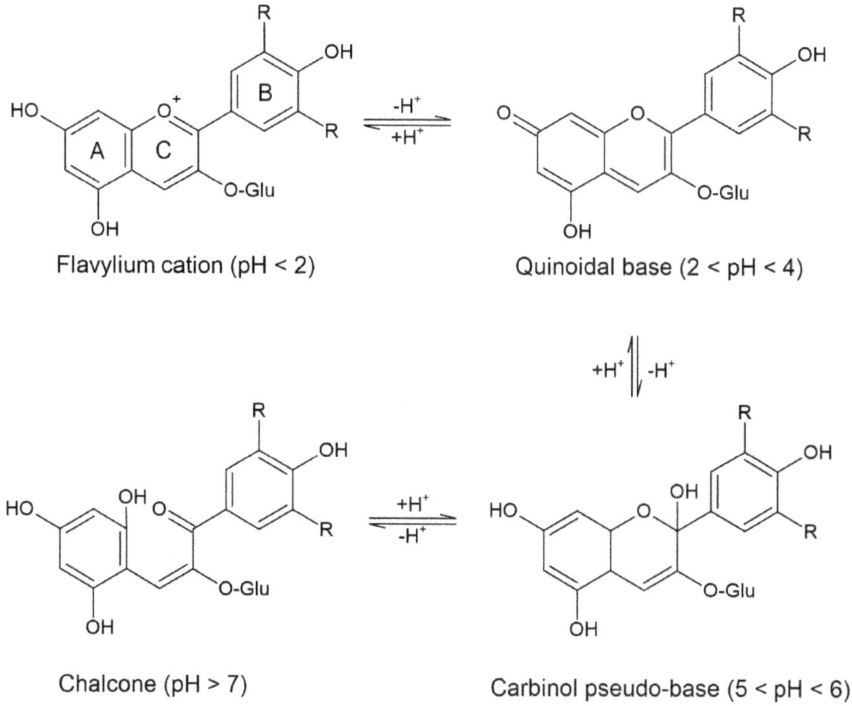

Scheme 3.1 Modification of chemical structures of the anthocyanins at different pH.

bluish coloration is obtained by anthocyanins linked to larger numbers of hydroxyl groups.[41]

Due to their ability to change color in relation to environmental conditions, anthocyanins are commonly used in the development of pH-sensitive indicators.[48,56,66,70] Anthocyanins are more stable under acidic conditions than under alkaline solutions, flavylium cations are formed in an acidic medium, giving the red-purple coloration.[71] With the increase in pH, anthocyanins present different types of stabilized structures, between pH values of 2 and 4, the pigment color can change to a bluish tone due to the predominance of hemiketal structures, such as those with anhydrous quinoid base.[41] pH values close to neutral (between 5 and 6) modify the structures of anthocyanins for colorless carbinol pseudobase.[72] When the medium becomes alkaline (pH > 7), there is an intense interaction between flavylium cations and chalcone caused by the nucleophilic reaction (Scheme 3.1), reducing the reddish color of anthocyanins and making them unstable and easily degraded due to disruption of the central ring (C), generating a yellowish-green color with the formation of chalcones.[41,66]

Furthermore, temperature is a factor that modifies the color of anthocyanins, enabling their use in the development of intelligent time–temperature indicators.[54] The action of temperature on anthocyanins occurs mainly by reducing the methoxy groups of the acetylated portion of anthocyanins, reducing their stability and causing the rupture of the central pyran ring and formation of yellowish chalcone.[68,73]

On the other hand, studies have reported the possibility of the application of anthocyanins as a natural dye to generate indications of freshness in protein-rich foods (*e.g.* milk, meat, yogurts and fish).[66,70,74,75] During the deterioration of protein-rich foods, volatile nitrogen compounds are released, which diffuse over the surface of intelligent films incorporated with anthocyanins, producing an alkaline environment due to the formation of hydroxyl ions.[74] Exposure of anthocyanins to this alkaline environment results in a change in the color of intelligent films (brown or yellow) due to chalcone formation.

3.3.2 Betalains

Betalains are water-soluble nitrogenous compounds, that are derived from betalamic acid and commonly found in the plant kingdom.[76] Betalain is an FDA-approved natural dye and its main sources are red pitaya, amaranth, beetroot and prickly pear.[46] The two major groups of betalains are betacyanin and betaxanthin (Scheme 3.2). Betacyanin shows a red-violet color due to the conjugation of betalamic acid with cyclo-Dopa, however, when beta-cyanine is linked to amino compounds (amines, glutamine, amino acids, *etc.*), they originate the yellow-orange betaxanthins.[68,76]

In general, betalains have been used as natural dyes and can also provide functional properties in foods as they demonstrate antimicrobial and antioxidant properties.[68,76] In addition, their application in the development of smart indicators has been studied.[6,24,35,77] Unlike anthocyanins, the betalains have greater colorimetric stability to pH change, being more stable at values between 3–7.[77]

Therefore, betalain is commonly used in the development of freshness indicators for meat products; the increase in volatile nitrogen content due to deterioration modifies the pigment structures, changing the initial red-violet

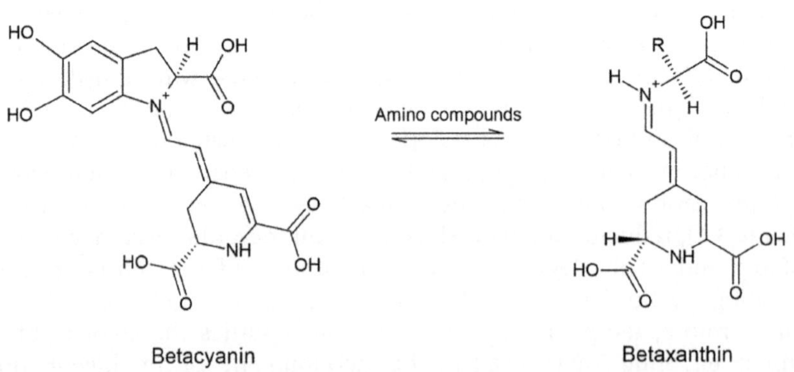

Scheme 3.2 Structure of the natural dyes betacyanin and betaxanthin.

color to yellow-orange.[35,77] The increase in the content of nitrogenous compounds promotes an alkaline environment, causing the cleavage of beta-cyanin and beta-cyanin cyclo-Dopa, resulting in a yellowish color.[6,68]

3.3.3 Carotenoids

Regarding lipophilic natural dyes, carotenoids are the most studied pigments.[68] Carotenoids are responsible for the yellow, orange and red colorings of plant foods (*e.g.* corn, pumpkin, carrots, papaya, tomato, *etc.*) and fish (*e.g.* salmon, shrimp, crab, *etc.*).[68,76] Carotenoids can be obtained from other unconventional sources such as insects, algae and micro-organisms.[41]

The carotenoids have structural diversity, being commonly composed of eight isoprenoid units and presenting a conjugated double bond in the central structure, acting as a chromophore.[68,76] This double bond is essential to perform its colorimetric and bioactive functions, since the action of this lipophilic pigment is associated with antioxidant and antimicrobial properties.[35,41,77] More than 1000 types of carotenoids are known,[78] with the most frequently found being β-carotene, lycopene, astaxanthin, lutein, zeaxanthin, bixin and norbixin,[76] as shown in Scheme 3.3.

As with other natural pigments, carotenoids are compounds susceptible to oxidation, isomerization and degradation by enzymatic processes and/or exposure to temperature, light, pH and O_2.[68] The main changes in carotenoid structure occur through geometric isomerization and oxidation of the compound, changing its intense yellow-orange color to pale yellow due to disruption of the central double bond and the loss of the chromophore molecule.[41] The application of carotenoids as a natural indicator pigment in intelligent packaging is associated with its susceptibility to color change due to exposure to inadequate storage temperature conditions, being commonly used in the evaluation of the quality and freshness of chilled fish or chicken.[55,77]

3.3.4 Chlorophylls

Chlorophylls are natural dyes responsible for the green color of plants and cyanobacteria. They are water-soluble compounds and their main structure is composed of porphyrin derivatives with a central metallic atom of magnesium and a hydrophobic phytol group.[41] In plants there are different types of chlorophyll (*e.g.* chlorophyll *a*, *b* or *c*), which can be identified according to their structure.[57] Chlorophyll *a* and *b* are the most common species, being identified by linkage with methyl (CH_3) or aldehyde (CHO) groups, respectively, in point (*R**) of Scheme 3.4.

Chlorophylls are highly susceptible to degradation by exposure to heat and acidic conditions, causing substitution of the magnesium molecule for a hydrogen molecule, producing light-yellow pheophytin.[41,79] Also, chlorophylls tend to be photo-oxidative and, when associated with carotenoids, can be used as a potential indicator of strong irradiation.[79] Changing color from exposure

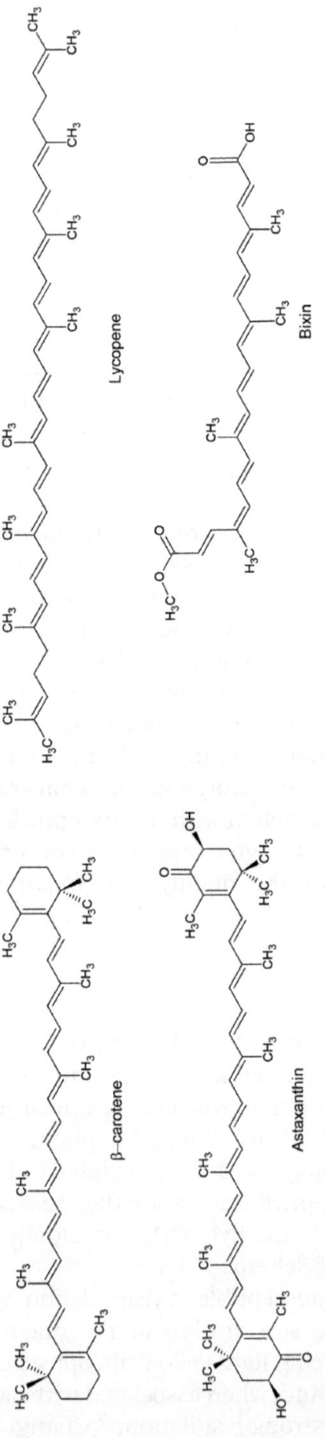

Scheme 3.3 Chemical structure of the main carotenoids found in nature.

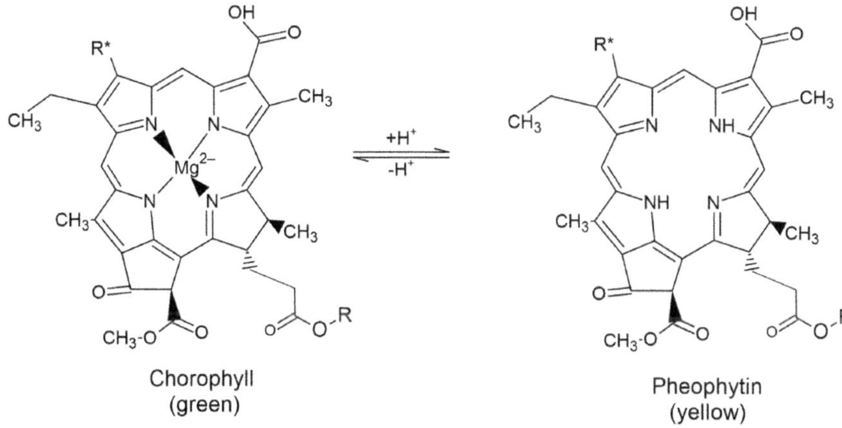

Scheme 3.4 Degradation of chlorophyll structure and formation of pheophytin under acidic conditions.

to adverse conditions makes chlorophylls a potential natural coloring indicator of quality in intelligent packaging, being able to assess the freshness and quality of sesame oil[57] and fish.[58]

3.3.5 Curcumin

Curcumin is extracted from the rhizomes of *Curcuma* sp., and is widely used in the preparation of condiments[61]. It is a di-phenolic dye with a hydrophobic characteristic,[46] which has excellent antioxidant, antimicrobial and anti-inflammatory properties.[80] Curcumin is formed by two *ortho*-methoxyphenol groups simultaneously linked by a β-diketone moiety (see Scheme 3.5).

The stability of the compound is pH dependent, in neutral to acidic conditions (pH < 7) the predominant structure is keto tautomerism, generating an intense yellow color.[81] When exposed to alkaline environments there is a predominance of the enol form of curcumin, represented by a dark red or brown coloration.[41] Therefore, curcumin has shown great potential in the application and development of colorimetric quality indicators, being able to detect factors of possible food degradation, such as pH change[60–62] and higher concentration of volatile amines.[59,82]

3.3.6 Phycocyanin

Phycocyanin is a natural water-soluble pigment belonging to the phycobilisome complex, obtained from prokaryotic cyanobacteria and eukaryotic red algae.[83] This pigment is formed by protein–pigment complexes and its structure is represented by a helix-shaped monomer, called α and β, linked to a bilin chromophore (see Scheme 3.6), responsible for the blue color of the molecule.[84]

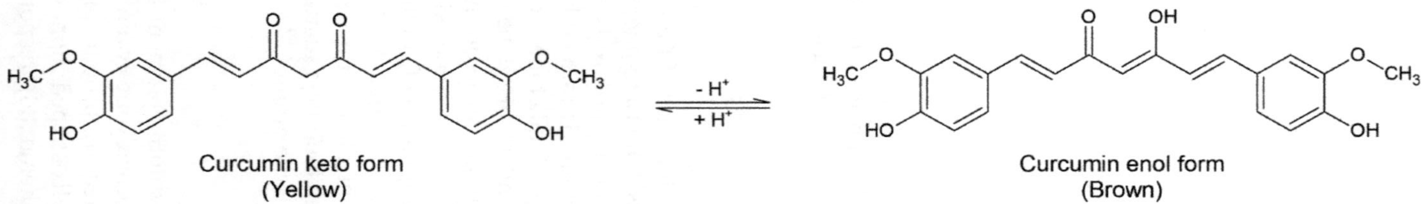

Scheme 3.5 Curcumin structure in acidic and alkaline conditions.

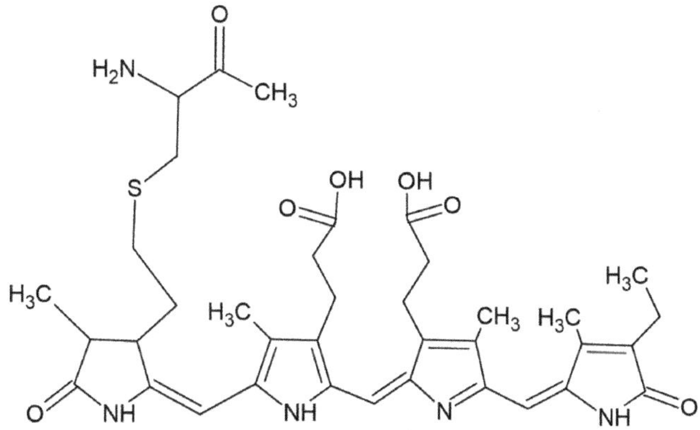

Scheme 3.6 Basic structure of the phycocyanin pigment.

The chromophore present in phycocyanins consists of an open-chain tetrapyrrole structure (phycobilin) covalently linked to protein molecules.[84] The stability and color loss of phycocyanins are related precisely to the modification of the chromophore structure,[62] therefore, any changes in the pigment–protein binding may favor the degradation and discoloration of the compound. The application of phycocyanins in the development of intelligent packaging remains quite limited, however, studies have shown that the incorporation of this pigment in biodegradable polymers can generate effective pH indicators capable of monitoring food quality.[62,63]

3.3.7 Quercetin

Belonging to the class of phenolic compounds, quercetin is a yellow-colored flavonoid,[62] widely known for its high antioxidant activity. Quercetin exhibits hydrophilic behavior and can be found in abundance in various plants (*e.g.* onions, apples, grapes, *etc.*) and beverages (*e.g.* tea, wine, *etc.*).[46] As it belongs to the flavonoid class, the structure of quercetins is similar to that of anthocyanins, however, the central ring is in the form of a pyrone, linked to a carbonyl group (–CO),[85] as shown in Scheme 3.7.

Exposure of this pigment in alkaline pH solutions ($pH > 7$) demonstrates a great influence on quercetin auto-oxidation, changing its color from yellow to red.[62] Furthermore, it is possible to observe color changes in relation to environmental factors (*e.g.* temperature and UV radiation), related to weathering and thermo-oxidation.[46,86] Therefore, quercetin has become a promising natural dye for the development of intelligent packaging that indicates aging time and food quality.

3.4 Potential Applications of Bio-based Indicators

Microbial spoilage and compound oxidation (*e.g.* flavors, lipid and vitamins oxidation) are among the main mechanisms of food degradation. The

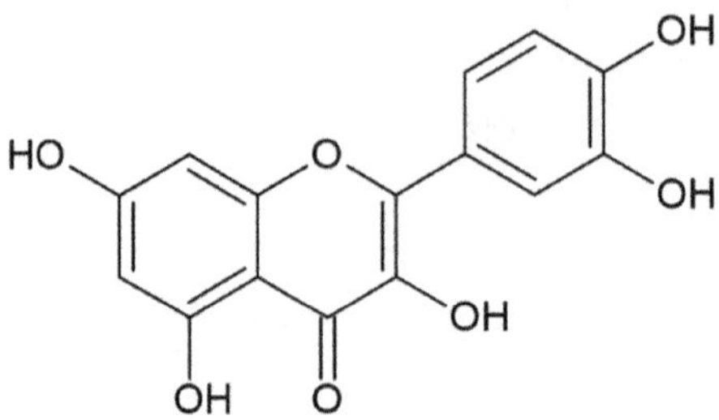

Scheme 3.7 Chemical structure of quercetin.

growth of deteriorating and pathogenic microorganisms in food products, as well as the development of oxidation reactions, result in changes in the pH and the production of compounds as organic acids, ethanol, biogenic amines, volatile nitrogen, and others. Such modifications might be measured by intelligent films produced with natural dyes to indicate the quality of food products.[87,88]

Based on the food degradation reactions, the quality of food products can be monitored by intelligent packaging using natural dyes. The sensors/indicators used in these types of packaging are generally of time–temperature, pH, gases and volatile compounds, freshness and integrity, chemicals, toxins and microorganisms.[41] Films incorporated with natural dyes have been applied in real food models, mostly in high protein content foods, such as meat products, milk, fruits and vegetables. Examples of studies in which natural dyes have been incorporated in films and applied to real foods are shown in Table 3.3.

Natural dyes such as anthocyanins, betalains, curcumin and others, obtained from various food-based sources including fruits, vegetables, flowers and herbs have been applied to different foods. Their utilization is advantageous because they are extracted from renewable sources which are widely available in nature. Most of the applications in foods refer to anthocyanins, followed by curcumin, probably because of its properties and easier obtainment. In the literature, studies have explored the incorporation of these natural dyes in various polymeric matrices, including synthetic (*e.g.* polyvinyl alcohol, polyethylene oxide, polyethylene, *etc.*) and natural polymers (*e.g.* starch, cellulose, chitosan, pectin, *etc.*). Biopolymers produced with polysaccharides, proteins and lipids incorporated with natural dyes to generate intelligent films contribute, besides the indication of spoilage or freshness of the products, to decreasing the use of conventional synthetic plastics, which are causing environmental problems.

The application of intelligent films produced with natural dyes has been explored mostly in meat products, fish and milk, due to their color changes at different conditions in terms of temperature, humidity, pH values, production of various gases, *etc.*[41] These products have high protein content, and the microbiological spoilage of protein produces elevated levels of total

Table 3.3 Applications of films containing natural dyes in real food products.

Natural dye	Dye source	Film structure	Foodstuff	References
Anthocyanins	Roselle calyx	Chitosan/ starch/ polyvinyl alcohol	Pork meat	Zhang *et al.* (2019)[74]
Anthocyanins	Roselle calyx	Polyvinyl alcohol/ sodium alginate	Pork griskin	Zhang *et al.* (2021)[89]
Anthocyanins	Beetroot	Pectin	Chilled beef	Guo *et al.* (2021)[90]
Anthocyanins	Purple tomato	Chitosan	Milk and fish	Li *et al.* (2021)[66]
Anthocyanins	Grape skins	Tara gum/ cellulose nanocrystals	Milk	Ma and Wang (2016)[91]
Anthocyanins	Black carrot	Cellulose/ chitosan	Milk	Ebrahimi Tirtashi *et al.* (2019)[92]
Anthocyanins	*Lycium ruthenicum* Murr (LRA)	Cassava starch	Freshness of pork/ shrimp and fish	Qin *et al.* (2019)[75]
Curcumin and anthocyanins	Purple sweet potato	Starch/ polyvinyl alcohol	Bighead carp fish	Chen *et al.* (2020)[93]
Anthocyanins and betalains	Beet root, elderberry, blueberry, green tea, and Yerba mate	Furcellaran gum	Fresh Atlantic mackerel/ seafood	Jamróz *et al.* (2019)[94]
Curcumin	—	Low-density polyethylene	Fresh beef and silver carp fish	Zhai *et al.* (2020)[59]
Curcumin	—	Chitosan and polyethylene oxide	Chicken	Yildiz *et al.* (2021)[82]
Curcumin	—	κ-Carrageenan gum	Pork and shrimp	Liu *et al.* (2018)[95]
Betalains	*Amaranthus* leaf extract	Polyvinyl alcohol and gelatin	Chicken/ fish	Kanatt (2020)[77]
Alizarin	—	Starch cellulose	Rainbow trout fish	Ezati *et al.* (2019)[96]

volatile basic nitrogen (TVB-N) containing ammonia, dimethylammonium and trimethylamine, which can modify the pH value of the medium.[95] Thus, this is an important mechanism used to indicate the product's quality using films containing natural dyes.

Anthocyanins obtained from roselle calyx, for example, were incorporated in films applied to pork meat[74] and pork griskin[89] (Table 3.3). Zhang *et al.*[74] reported that the intelligent films also presented antioxidant activity besides color changing. Additionally, they had better color stability and improved the tensile strength of films. The roselle anthocyanin color changed from red to blue then to yellow when the pH increased from 2 to 12. It became red when the pH was no more than 5.0, pink at pH 5–6, purple at pH 8–9, and yellow at pH 11–12. The starch/polyvinyl alcohol films incorporated with anthocyanin film showed visible changes from red to green when employed to monitor the freshness of pork stored at 25 °C, before the TVB-N value of the pork gradually increased to the rejection limit (15 mg/100 g) at 36 h. Zhang *et al.*[89] developed intelligent indicator films incorporating roselle anthocyanins in polyvinyl alcohol/sodium alginate films to monitor griskin freshness. The TVB-N level increased to 18.02 mg/100 g in 72 h, and the color of the indicator also presented visible changes. The results of this study indicated a good correlation between TVB-N and pH with color change of the indicator.

Anthocyanins were also studied to indicate changes in milk (Table 3.3). Ma and Wang[91] incorporated extracts from grape skins into tara gum/cellulose nanocrystal matrix to prepare a colorimetric pH-sensing film. The extract affected the film's structure by decreasing the compact and continuous structure, thermal stability, tensile strength and barrier properties of the films, while elongation at break and the oxygen permeability of films increased significantly. Regarding the color of films, it varied from red (in acid pH) to slightly green (in alkali pH) (pH 1–11). In the evaluation of milk microbial contamination and consequently the pH modification, in the unspoiled milk the film showed a yellow color. After 48 h, the film became slightly darker and redder, clearly indicating that the milk had spoiled, and the pH decreased. Ebrahimi Tirtashi *et al.*[92] used anthocyanins from black carrot as a chemo-responsive dye to produce a colorimetric pH indicator in a cellulose–chitosan matrix to monitor spoilage in pasteurized milk. The colorimetric pH indicator showed color variation at different pH values (pH 2–11). In the fresh pasteurized milk test, a perceptible color change from blue to violet rose color after 48 h of storage at 20 °C was easily observed by eye. Thus, this indicator can be used as a food-grade biomaterial to monitor the freshness/spoilage of milk. Additionally, stability tests revealed that the indicator had color stability during 1-month storage at 20 °C.

Curcumin also showed potential for use in intelligent indicators for beef, chicken, fish, pork and shrimp (Table 3.3). Zhai *et al.*[59] developed a film using low-density polyethylene (LDPE) and curcumin with the melting extrusion method. The LDPE–curcumin composite film was sensitive to ammonia (NH_3). The LDPE–curcumin film produced was used to monitor beef and silver carp spoilage at 4 °C. Color changes were observed from light yellow to light brown due to the increase in TVB-N contents of the samples. Although the polymer matrix used is not sustainable, this film is especially advantageous because it is suitable for industrial production on a large scale. Liu *et al.*[95] also used curcumin, but in a bio-based matrix (κ-carrageenan). Besides the functionalities provided by curcumin, its incorporation significantly improved the film barrier and tensile strength and enhanced thermal stability. The produced films showed a strong redness-shift under alkaline conditions, from yellow to red, and successfully indicated pH modifications of pork and shrimp. These results indicated that these films may also have strong potential to monitor the freshness of other animal-based protein foods.

Kanatt[77] incorporated *Amaranthus* leaf extract in a film developed with polyvinyl alcohol (PVA) and gelatin to monitor the freshness of fish and chicken meat. *Amaranthus* is rich in phenolic compounds and betalains, which provide active and intelligent properties to the film. The incorporation of *Amaranthus* leaf extract decreased water solubility and water vapor permeability and increased mechanical properties. In the evaluation of chilled fish/chicken, the films were active in reducing oxidative rancidity and delaying microbial growth. Regarding the intelligent properties, it showed a visible color shift from red to yellow that resulted from spoilage, which agreed with increases in pH, TVB-N and microbial counts. Therefore, the PVA–gelatin film incorporated with *Amaranthus* leaf extract could be used as active and intelligent food packaging.

Alizarin is another natural dye that exhibits color changes at different pH, varying from yellow in acidic conditions to red in basic conditions. Ezati *et al.*[96] studied the incorporation of alizarin in starch–cellulose films to track the freshness of rainbow trout fillet. They evaluated fish spoilage by sticking the indicator on the headspace of a sterilized packaging box containing fish fillet. The pH, microbiological analysis and TVB-N determinations were performed. The freshness indicators showed color changes from yellow to purple and dark red at pH 2–11 and high sensitivity and color response to ammonia. Changes at pH 7–9 were detectable by eye. Thus, the alizarin pH indicator produced with starch and cellulose can monitor the freshness of fish products.

Natural dyes from different sources can be used individually, as discussed above, or combined in a film to generate intelligent indicators, such as curcumin + anthocyanins in the study of Chen *et al.*[93] These authors studied curcumin, anthocyanins, and their mixture incorporated in starch/polyvinyl alcohol films. The films were used as on-package labels (placed inside the packaging without contact with the sample) to monitor fish freshness by responding to the increase of volatile amine release. The film with anthocyanins showed the worst color stability, while the film containing curcumin had the best, however all films showing a color change rate have potential applications. The film incorporated with the mixture of curcumin + anthocyanins in the ratio of 2 : 8 (v/v) revealed the greatest correlation of color change to the degree of fish freshness. It presented three different colors which were equivalent to the freshness, medium freshness and spoilage of fish. Such changes could be recognized by eye indicating that this is a potential indicator material to nondestructively monitor fish freshness.

Exploring the studies reported in the literature, it is possible to observe that intelligent films produced with natural dyes have great potential applications to monitor/indicate food freshness and spoilage. The use of natural dyes in synthetic polymers is advantageous due to the possibility to produce them on a large scale easily, while their use in biopolymers is positive in terms of sustainability. Regardless of the type of the polymer, the natural dyes are effective to monitor/indicate reactions based on the degradation mechanisms of different foods. Besides that, most natural dyes also have the capacity to prevent degradation reactions due to their active nature (antioxidant and antimicrobial properties) and they might contribute to an increase in the packaging performance (*e.g.* barrier, strength) of biopolymers. The combination of more than one natural dye in a film seems to collaborate in the production of improved performance indicators, in terms of correlation between the color changing and the product characteristics as well as the color stability of the indicator.

3.5 Final Considerations

Over the years, concerns about ensuring the quality and safety of food have boosted the development of intelligent packaging, such as freshness indicators. Studies have shown that natural dyes are a non-toxic, biocompatible and hypoallergenic alternative for the development of quality indicators for packaged foods. The main source of natural dyes is food waste (*e.g.* peel, leaves, skins, *etc.*), being a sustainable and low-cost option to produce smart indicators. Studies have shown that freshness indicators developed with natural dyes are highly efficient in monitoring the quality of protein-rich products (*e.g.* meat, chicken, fish, milk, *etc.*). However, further studies are needed in relation to the application of these freshness indicators in a greater variety of foods, showing their real application potential in indicating the quality of various packaged foods, encouraging their application in different types of marketed products.

References

1. Y. Zhang, Q. Zhou, L. J. Rather and Q. Li, Agricultural waste of *Eriobotrya japonica* L. (Loquat) seeds and flora leaves as source of natural dye and bio-mordant for coloration and bio-functional finishing of wool textile, *Ind. Crops Prod.*, 2021, **169**, 113633.
2. K. Phan, K. Raes, V. V. Speybroeck, M. Roosen, K. Clerck and S. Meester, Non-food applications of natural dyes extracted from agro-food residues: A critical review, *J. Cleaner Prod.*, 2021, **301**, 126920.
3. S. Ribeiro and C. M. Veloso, Microencapsulation of natural dyes with biopolymers for application in food: A review, *Food Hydrocolloids*, 2021, **112**, 106374.

4. J. L. D. Antigo, A. P. Stafussa, R. D. C. Bergamasco and G. S. Madrona, Chia seed mucilage as a potential encapsulating agent of a natural food dye, *J. Food Eng.*, 2020, **285**, 110101.
5. A. Fajraoui, J. B. Nasr, C. Lacoste, M. B. Amar, P. Dony, S. Odof and F. E. L. Halouani, Coloration of the polylactic acid with the natural dye extracted from acacia cyanophylla flowers, *Polym. Test.*, 2019, **78**, 105988.
6. Y. Qin, Y. Liu, X. Zhang and J. Liu, Development of active and intelligent packaging by incorporating betalains from red pitaya (*Hylocereus polyrhizus*) peel into starch/polyvinyl alcohol films, *Food Hydrocolloids*, 2020, **100**, 105410.
7. G. Cuoco, C. Mathe, P. Archier and C. Vieillescazes, Characterization of madder and garancine in historic French red materials by liquid chromatography-photodiode array detection, *J. Cult. Herit.*, 2011, **12**, 98–104.
8. K. Mayolo-Deloisa, J. Rodríguez-Rodríguez, G. Velázquez, K. Mayolo-Deloisa, M. Rito-Palomares, J. A. Torres and R. Parra-Saldívar, Low-sugar content betaxanthins extracts from yellow pitaya (*Stenocereus pruinosus*), *Food Bioprod. Process.*, 2020, **121**, 178–185.
9. A. Stupar, V. Seregelj, B. Dias, L. Pezo, A. Cvetanovi, A. Mi and I. Marrucho, Recovery of β-carotene from pumpkin using switchable natural deep eutectic solvents, *Ultrason. Sonochem.*, 2021, **76**, 105638.
10. J.-H. Chen, Y. Kato, M. Matsuda, C.-Y. Chen, D. Nagarajan, T. Hasunuma, A. Kondo and J.-S. Chang, Lutein production with chlorella sorokiniana MB-1-M12 using novel two-stage cultivation strategies – metabolic analysis and process improvement, *Bioresour. Technol.*, 2021, **334**, 125200.
11. J. Dhakane-Lad and A. Kar, Supercritical CO_2 extraction of lycopene from pink grapefruit (*Citrus paradise* Macfad) and its degradation studies during storage, *Food Chem.*, 2021, **361**, 130113.
12. D. Jiménez, R. Vardanega, F. Salinas, C. Espinosa-álvarez, W. Bugueño-muñoz, J. Palma, M. A. A. Meireles and M. C. Ruíz-domínguez, Effect of drying methods on biorefinery process to obtain capsanthin and phenolic compounds from Capsicum annuum L, *J. Supercrit. Fluids.*, 2021, **174**, 105241.
13. L. Wang, J. Hu, W. Lv, W. Lu, D. Pei, Y. Lv, W. Wang, M. Zhang, R. Ding and M. Lv, Optimized extraction of astaxanthin from shrimp shells treated by biological enzyme and its separation and purification using macroporous resin, *Food Chem.*, 2021, **363**, 130369.
14. S. S. Abuthahir, C. K. Venil, M. Malathi and P. R. Devi, Optimization of submerged fermentation for enhanced production of canthaxanthin by Dietzia maris AURCCBT01, *Mater. Today: Proc.*, 2021, **47**, 2132–2137.
15. L. Bourdon, A. A. Jensen, J. M. Kavanagh and D. D. Mcclure, Microalgal production of zeaxanthin, *Algal Res.*, 2021, **55**, 102266.
16. G. Petruk, A. Illiano, R. Del, A. Raiola, A. Amoresano, M. Manuela, R. Piccoli and D. Maria, Malvidin and cyanidin derivatives from açai fruit (*Euterpe oleracea* Mart.) counteract UV-A-induced oxidative

stress in immortalized fibroblasts, *J. Photochem. Photobiol., B*, 2017, **172**, 42–51.
17. V. K. Rajan, C. K. Hasna and K. Muraleedharan, The natural food colorant Peonidin from cranberries as a potential radical scavenger – A DFT based mechanistic analysis, *Food Chem.*, 2018, **262**, 184–190.
18. J. Zhang, Y. Ge, P. Liu, D. Wu and H. Liu, Biotechnological strategies of riboflavin biosynthesis in microbes, *Engineering*, 2021, **12**, 115–127.
19. N. Aliabbasi, M. Fathi and Z. Emam-djomeh, Curcumin: A promising bioactive agent for application in food packaging systems, *J. Environ. Chem. Eng.*, 2021, **9**, 105520.
20. Z. Mousavi, M. H. M. Ara, B. Ghafary and I. Ahanian, Magneto-optical properties of olive oil combined with natural turmeric powder as a dye, *Opt. Laser. Technol.*, 2021, **137**, 106801.
21. S. C. Ezike, C. N. Hyelnasinyi, M. A. Salawu, J. F. Wansah, A. N. Ossai and N. N. Agu, Synergestic effect of chlorophyll and anthocyanin Co-sensitizers in TiO_2-based dye-sensitized solar cells, *Surf. Interfaces*, 2021, **22**, 100882.
22. Y. Caro, L. Anamale, M. Fouillaud, P. Laurent, T. Petit and L. Dufosse, Natural hydroxyanthraquinoid pigments as potent food grade colorants: an overview, *Nat. Prod. Bioprospect.*, 2012, **2**, 174–193.
23. E. G. D. Mejia, Q. Zhang, K. Penta, A. Eroglu and M. A. Lila, The Colors of Health: Chemistry, Bioactivity, and Market Demand for Colorful Foods and Natural Food Sources of Colorants, *Annu. Rev. Food Sci. Technol.*, 2020, **30**, 1–38.
24. W. Yang, M. Kaimainen, E. Järvenpää, M. Sandell, R. Huopalahti, B. Yang and O. Laaksonen, Red beet (*Beta vulgaris*) betalains and grape (*Vitis vinifera*) anthocyanins as colorants in white currant juice – Effect of storage on degradation kinetics, color stability and sensory properties, *Food Chem.*, 2021, **348**, 128995.
25. Y. Ju, L. Yang, X. Yue, Y. Li, R. He, S. Deng, X. Yang and Y. Fang, Anthocyanin profiles and color properties of red wines made from Vitis davidii and Vitis vinifera grapes, *Food Sci. Hum. Wellness*, 2021, **10**, 335–344.
26. B. R. Albuquerque, J. Pinela, L. Barros, M. B. P. P. Oliveira and I. C. F. R. Ferreira, Anthocyanin-rich extract of jabuticaba epicarp as a natural colorant: Optimization of heat- and ultrasound-assisted extractions and application in a bakery product, *Food Chem.*, 2020, **316**, 126364.
27. A. B. Das, V. V. Goud and C. Das, Microencapsulation of anthocyanin extract from purple rice bran using modified rice starch and its effect on rice dough rheology, *Int. J. Biol. Macromol.*, 2019, **124**, 573–581.
28. E. Pieczykolan and M. A. Kurek, Use of guar gum, gum arabic, pectin, beta-glucan and inulin for microencapsulation of anthocyanins from chokeberry, *Int. J. Biol. Macromol.*, 2019, **129**, 665–671.
29. W. Quan, W. He, X. Qie, Y. Chen, M. Zeng, F. Qin, J. Chen and Z. He, Effects of β-cyclodextrin, whey protein, and soy protein on the thermal

and storage stability of anthocyanins obtained from purple-fleshed sweet potatoes, *Food Chem.*, 2020, **320**, 126655.
30. L. G. Santos, G. F. A. Silva, B. M. Gomes and V. G. Martins, A novel sodium alginate active films functionalized with purple onion peel extract (*Allium cepa*), *Biocatal. Agric. Biotechnol.*, 2021, **35**, 102096.
31. Z. Chai, D. D. Herrera-balandrano, H. Yu, T. Beta, Q. Zeng, X. Zhanga, L. Tian, L. Niu and W. Huang, A comparative analysis on the anthocyanin composition of 74 blueberry cultivars from China, *J. Food Compos. Anal.*, 2021, **102**, 104051.
32. A. P. Preczenhak, B. Orsi, G. P. P. Lima, J. V. Tezotto-Uliana, I. O. Minatel and R. A. Klug, Cysteine enhances the content of betalains and polyphenols in fresh-cut red beet, *Food Chem.*, 2019, **286**, 600–607.
33. M. J. Cejudo-Bastante, N. Hurtado, N. Mosquera and F. J. Heredia, Potential use of new Colombian sources of betalains. Color stability of ulluco (*Ullucus tuberosus*) extracts under different pH and thermal conditions, *Food Res. Int.*, 2014, **64**, 465–471.
34. M. J. Cejudo-Bastante, N. Hurtado and F. J. Heredia, Potential use of new Colombian sources of betalains. Colorimetric study of red prickly pear (*Opuntia dillenii*) extracts under different technological conditions, *Food Res. Int.*, 2015, **71**, 91–99.
35. H. Hu, X. Yao, Y. Qin, H. Yong and J. Liu, Development of multifunctional food packaging by incorporating betalains from vegetable amaranth (*Amaranthus tricolor* L.) into quaternary ammonium chitosan/fish gelatin blend films, *Int. J. Biol. Macromol.*, 2020, **159**, 675–684.
36. M. Montiel-Sánchez, T. García-Cayuela, A. Gómez-Maqueo and H. S. García, In vitro gastrointestinal stability, bioaccessibility and potential biological activities of betalains and phenolic compounds in cactus berry fruits (*Myrtillocactus geometrizans*), *Food Chem.*, 2021, **342**, 128087.
37. K. Sinha, S. Chowdhury, P. Das and S. Datta, Modeling of microwave-assisted extraction of natural dye from seeds of *Bixa orellana* (Annatto) using response surface methodology (RSM) and artificial neural network (ANN), *Ind. Crop. Prod.*, 2013, **41**, 165–171.
38. P. M. Lima, G. C. Dacanal, L. S. Pinho, L. J. Pérez-Córdoba, M. Thomazini, I. C. F. Moraes and C. S. Favaro-Trindade, Production of a rich-carotenoid colorant from pumpkin peels using oil-in-water emulsion followed by spray drying, *Food. Res. Int.*, 2021, **148**, 110627.
39. L. E. Ordóñez-Santos, J. Esparza-Estrada and P. Vanegas-Mahecha, Ultrasound-assisted extraction of total carotenoids from mandarin epicarp and application as natural colorant in bakery products, *LWT*, 2021, **139**, 110598.
40. L. Gui, L. Xu, Z. Liu, Z. Zhou and Z. Sun, Carotenoid-rich microalgae promote growth and health conditions of *Artemia nauplii*, *Aquaculture*, 2021, 737289.
41. M. Alizadeh-Sani, E. Mohammadian, J. W. Rhim and S. M. Jafari, pH-sensitive (halochromic) smart packaging films based on natural food

colorants for the monitoring of food quality and safety, *Trends. Food Sci. Technol.*, 2020, **105**, 93–144.
42. E. Balbinot-Alfaro, D. V. Craveiro, K. O. Lima, H. L. G. Costa, D. R. Lopes and C. Prentice, Intelligent packaging with pH indicator potential, *Food Eng. Rev.*, 2019, **11**, 235–244.
43. C. Rukchon, A. Nopwinyuwong, S. Trevanich, T. Jinkarn and P. Suppakul, Development of a food spoilage indicator for monitoring freshness of skinless chicken breast, *Talanta*, 2014, **130**, 547–554.
44. H. Z. Chen, M. Zhang, B. Bhandari and Z. Guo, Applicability of a colorimetric indicator label for monitoring freshness of fresh-cut green bell pepper, *Postharvest. Biol. Technol.*, 2018, **140**, 85–92.
45. A. Soni, Development of colorimetric on-package indicator for monitoring of chicken meat freshness during refrigerated storage (4 ± 1 °C), *J. Anim. Res.*, 2018, **8**, 847–853.
46. N. Bhargava, V. S. Sharanagat, R. S. Mor and K. Kumar, Active and intelligent biodegradable packaging films using food and food waste-derived bioactive compounds: A review, *Trends. Food Sci. Technol.*, 2020, **105**, 385–401.
47. Y. S. Musso, P. R. Salgado and A. N. Mauri, Smart gelatin films prepared using red cabbage (*Brassica oleracea* L.) extracts as solvent, *Food Hydrocolloids*, 2019, **89**, 674–681.
48. G. S. Filipini, V. P. Romani and V. G. Martins, Biodegradable and active-intelligent films based on methylcellulose and jambolão (*Syzygium cumini*) skins extract for food packaging, *Food Hydrocolloids*, 2020, **109**, 106139.
49. S. G. Kuntzler, J. A. V. Costa, A. P. D. R. Brizio and M. G. D. Morais, Development of a colorimetric pH indicator using nanofibers containing *Spirulina* sp. LEB 18, *Food Chem.*, 2020, **328**, 126768.
50. M. S. Firouz, K. Mohi-Alden and M. Omid, A critical review on intelligent and active packaging in the food industry: Research and development, *Food Res. Int.*, 2021, **141**, 110113.
51. T. Liang, G. Sun, L. Cao, J. Li and L. Wang, Rheological behavior of film-forming solutions and film properties from Artemisia sphaerocephala Krasch. gum and purple onion peel extract, *Food Hydrocolloids*, 2018, **82**, 124–134.
52. J. Liu, H. Wang, M. Guo, L. Li, M. Chen, S. Jiang, X. Li and S. Jiang, Extract from Lycium ruthenicum Murr. Incorporating κ-carrageenan colorimetric film with a wide pH–sensing range for food freshness monitoring, *Food Hydrocolloids*, 2019, **94**, 1–10.
53. V. A. Pereira, I. N. Q. Arruda and R. Stefani, Active chitosan/PVA films with anthocyanins from Brassica oleraceae (Red Cabbage) as Time–Temperature Indicators for application in intelligent food packaging, *Food Hydrocolloids*, 2015, **43**, 180–188.
54. V. B. V. MacIel, C. M. P. Yoshida and T. T. Franco, Development of a prototype of a colourimetric temperature indicator for monitoring food quality, *J. Food Eng.*, 2012, **111**, 21–27.

55. E. Jamróz, P. Kulawik, P. Krzyściak, K. Talaga-Ćwiertnia and L. Juszczak, Intelligent and active furcellaran-gelatin films containing green or pu-erh tea extracts: Characterization, antioxidant and antimicrobial potential, *Int. J. Biol. Macromol.*, 2019, **122**, 745–757.
56. M. Latos-Brozio and A. Masek, The application of natural food colorants as indicator substances in intelligent biodegradable packaging materials, *Food Chem. Toxicol.*, 2020, **135**, 110975.
57. S. Chavoshizadeh, S. Pirsa and F. Mohtarami, Sesame oil oxidation control by active and smart packaging system using wheat gluten/chlorophyll film to increase shelf life and detecting expiration date, *Eur. J. Lipid. Sci. Technol.*, 2020, **122**, 1–39.
58. S. Jancikova, E. Jamróz, P. Kulawik, J. Tkaczewska and D. Dordevic, Furcellaran/gelatin hydrolysate/rosemary extract composite films as active and intelligent packaging materials, *Int. J. Biol. Macromol.*, 2019, **131**, 19–28.
59. X. Zhai, X. Wang, J. Zhang, Z. Yang, Y. Sun, Z. Li, X. Huang, M. Holmes, Y. Gong, M. Povey, J. Shi and X. Zou, Extruded low density polyethylene-curcumin film: A hydrophobic ammonia sensor for intelligent food packaging, *Food Packag. Shelf Life*, 2020, **26**, 100595.
60. A. Etxabide, P. A. Kilmartin and J. I. Maté, Color stability and pH-indicator ability of curcumin, anthocyanin and betanin containing colorants under different storage conditions for intelligent packaging development, *Food Control.*, 2021, **121**, 107645.
61. Y. S. Musso, P. R. Salgado and A. N. Mauri, Smart edible films based on gelatin and curcumin, *Food Hydrocolloids*, 2017, **66**, 8–15.
62. A. L. M. Terra, J. B. Moreira, J. A. V. Costa and M. G. D. Morais, Development of time-pH indicator nanofibers from natural pigments: An emerging processing technology to monitor the quality of foods, *LWT*, 2021, **142**, 111020.
63. J. B. Moreira, A. L. M. Terra, J. A. V. Costa and M. G. D. Morais, Development of pH indicator from PLA/PEO ultrafine fibers containing pigment of microalgae origin, *Int. J. Biol. Macromol.*, 2018, **118**, 1855–1862.
64. R. Bai, X. Zhang, H. Yong, X. Wang, Y. Liu and J. Liu, Development and characterization of antioxidant active packaging and intelligent Al^{3+}-sensing films based on carboxymethyl chitosan and quercetin, *Int. J. Biol. Macromol.*, 2019, **126**, 1074–1084.
65. G. F. A. Silva, L. G. Santos, B. M. Gomes and V. Guimarães Martins, Clitoria ternatea: Avaliação do tempo de extração de compostos fenólicos obtidos pela técnica assistida por ultrassom, *Rev. Bras. Agrotecnol.*, 2021, **11**, 990–994.
66. Y. Li, K. Wu, B. Wang and X. Li, Colorimetric indicator based on purple tomato anthocyanins and chitosan for application in intelligent packaging, *Int. J. Biol. Macromol.*, 2021, **174**, 370–376.
67. S. Kalpana, S. R. Priyadarshini, M. Maria Leena, J. A. Moses and C. Anandharamakrishnan, Intelligent packaging: Trends and applications in food systems, *Trends. Food Sci. Technol.*, 2019, **93**, 145–157.

68. D. B. Rodriguez-Amaya, Update on natural food pigments – A mini-review on carotenoids, anthocyanins, and betalains, *Food Res. Int.*, 2019, **124**, 200–205.
69. H. Yong and J. Liu, Recent advances in the preparation, physical and functional properties, and applications of anthocyanins-based active and intelligent packaging films, *Food Packag. Shelf Life*, 2020, **26**, 100550.
70. S. Mohammadalinejhad, H. Almasi and M. Moradi, Immobilization of *Echium amoenum* anthocyanins into bacterial cellulose film: A novel colorimetric pH indicator for freshness/spoilage monitoring of shrimp, *Food Control.*, 2020, **113**, 107169.
71. M. M. Giusti and G. T. Sigurdson, *Advances in Anthocyanin Research 2018*, MDPI, 2019.
72. W. G. Sganzerla, C. P. Pereira Ribeiro, N. R. Uliana, M. B. Cassetari Rodrigues, C. G. D. Rosa, J. P. Ferrareze, A. P. D. L. Veeck and M. R. Nunes, Bioactive and pH-sensitive films based on carboxymethyl cellulose and blackberry (*Morus nigra* L.) anthocyanin-rich extract: A perspective coating material to improve the shelf life of cherry tomato (*Solanum lycopersicum* L. var. cerasiforme), *Biocatal. Agric. Biotechnol.*, 2021, **33**, 101989.
73. Y. Bi, X. Chi, R. Zhang, Y. Lu, Z. Wang, Q. Dong, C. Ding, R. Yang and L. Jiang, Highly efficient extraction of mulberry anthocyanins in deep eutectic solvents: Insights of degradation kinetics and stability evaluation, *Innov. Food Sci. Emerg. Technol.*, 2020, **66**, 102512.
74. J. Zhang, X. Zou, X. Zhai, X. Huang, C. Jiang and M. Holmes, Preparation of an intelligent pH film based on biodegradable polymers and roselle anthocyanins for monitoring pork freshness, *Food Chem.*, 2018, **272**, 306–312.
75. Y. Qin, Y. Liu, H. Yong, J. Liu, X. Zhang and J. Liu, Preparation and characterization of active and intelligent packaging films based on cassava starch and anthocyanins from *Lycium ruthenicum* Murr, *Int. J. Biol. Macromol.*, 2019, **134**, 80–90.
76. D. B. Rodriguez-Amaya, Natural food pigments and colorants, *Curr. Opin. Food Sci.*, 2016, 7, 20–26.
77. S. R. Kanatt, Development of active/intelligent food packaging film containing Amaranthus leaf extract for shelf life extension of chicken/fish during chilled storage, *Food Packag. Shelf Life*, 2020, **24**, 100506.
78. L. G. Santos and L. O. Werle, Influência da temperatura de secagem de fatias da polpa de abóbora menina brasileira nos parâmetros cinéticos e termodinâmicos, *Rev. Bras. Agrotecnol.*, 2021, **11**, 556–562.
79. N. M. Streit, L. P. Canterle, M. W. D Canto and L. H. H. Hecktheuer, As clorofilas, *Ciência Rural*, 2005, **35**, 748–755.
80. T. Jiang, R. Ghosh and C. Charcosset, Extraction, purification and applications of curcumin from plant materials-A comprehensive review, *Trends. Food Sci. Technol.*, 2021, **112**, 419–430.

81. Z. Liu, J. D. Smart and A. S. Pannala, Recent developments in formulation design for improving oral bioavailability of curcumin: A review, *J. Drug Delivery Sci. Technol.*, 2020, **60**, 102082.
82. E. Yildiz, G. Sumnu and L. N. Kahyaoglu, Monitoring freshness of chicken breast by using natural halochromic curcumin loaded chitosan/PEO nanofibers as an intelligent package, *Int. J. Biol. Macromol.*, 2021, **170**, 437–446.
83. G. Ferraro, P. Imbimbo, A. Marseglia, R. Lucignano, D. M. Monti and A. Merlino, X-ray structure of C-phycocyanin from galdieria phlegrea: Determinants of thermostability and comparison with a C-phycocyanin in the entire phycobilisome, *Biochim. Biophys. Acta, Bioenerg.*, 2020, **1861**, 148236.
84. D. Pez Jaeschke, I. Rocha Teixeira, L. Damasceno Ferreira Marczak and G. Domeneghini Mercali, Phycocyanin from spirulina: A review of extraction methods and stability, *Food Res. Int.*, 2021, **143**, 110314.
85. P. Singh, Y. Arif, A. Bajguz and S. Hayat, The role of quercetin in plants, *Plant. Physiol. Biochem.*, 2021, **166**, 10–19.
86. A. Masek, M. Latos, M. Piotrowska and M. Zaborski, The potential of quercetin as an effective natural antioxidant and indicator for packaging materials, *Food Packag. Shelf Life*, 2018, **16**, 51–58.
87. G. L. Robertson, in *Food packaging and shelf life: A practical guide*, ed. G. L. Robertson, CRC Press, Boca Raton, FL, 2009.
88. J. H. Han, C. H. L. Ho and E. T. Rodrigues Intelligent packaging. in *Innovations in Food Packaging*, ed. J. Han, Elsevier, 2005, pp. 138–155.
89. J. Zhang, X. Huang, J. Shi, L. Liu, X. Zhang, X. Zou, J. Xiao, X. Zhai, D. Zhang, Y. Li and T. Shen, A visual bi-layer indicator based on roselle anthocyanins with high hydrophobic property for monitoring griskin freshness, *Food Chem.*, 2021, **355**, 129573.
90. Z. Guo, X. Ge, W. Li, L. Yang, L. Han and Q. Yu, Active-intelligent film based on pectin from watermelon peel containing beetroot extract to monitor the freshness of packaged chilled beef, *Food Hydrocolloids*, 2021, **119**, 106751.
91. Q. Ma and L. Wang, Preparation of a visual pH-sensing film based on tara gum incorporating cellulose and extracts from grape skins, *Sens. Actuators, B*, 2016, **235**, 401–407.
92. F. Ebrahimi Tirtashi, M. Moradi, H. Tajik, M. Forough, P. Ezati and B. Kuswandi, Cellulose/chitosan pH-responsive indicator incorporated with carrot anthocyanins for intelligent food packaging, *Int. J. Biol. Macromol.*, 2019, **136**, 920–926.
93. H. Z. Chen, M. Zhang, B. Bhandari and C. H. Yang, Novel pH-sensitive films containing curcumin and anthocyanins to monitor fish freshness, *Food Hydrocolloids*, 2020, **100**, 105438.
94. E. Jamróz, P. Kulawik, P. Guzik and I. Duda, The verification of intelligent properties of furcellaran films with plant extracts on the stored fresh Atlantic mackerel during storage at 2 °C, *Food Hydrocolloids*, 2019, **97**, 105211.

95. J. Liu, H. Wang, P. Wang, M. Guo, S. Jiang, X. Li and S. Jiang, Films based on κ-carrageenan incorporated with curcumin for freshness monitoring, *Food Hydrocolloids*, 2018, **83**, 134–142.
96. P. Ezati, H. Tajik, M. Moradi and R. Molaei, Intelligent pH-sensitive indicator based on starch-cellulose and alizarin dye to track freshness of rainbow trout fillet, *Int. J. Biol. Macromol.*, 2019, **132**, 157–165.

CHAPTER 4

Organic–Inorganic Hybrid Materials for Active Packaging Applications

P. WOJCIECHOWSKA

Department of Industrial Products and Packaging Quality, Poznań University of Economics and Business, al. Niepodległości 10, 61-875 Poznań, Poland
Email: patrycja.wojciechowska@ue.poznan.pl

4.1 Organic–Inorganic Hybrid Materials

Organic–inorganic hybrids are materials consisting of two or more components, usually organic and inorganic ones, combined at a molecular level. According to the IUPAC (International Union of Pure and Applied Chemistry) definition: "A hybrid material is composed of an intimate mixture of inorganic components, organic components, or both types of components. The components usually interpenetrate on a scale of less than 1 μm".[1] It is worth noting that, in this perspective, this definition may include nanocomposites, since these materials contain, in an unbound state or as an aggregate, particles where one or more external dimensions is in the size range 1–100 nm.[2] Moreover, in the scientific literature, there is no sharp distinction between these two terms.[3] However, according to Kickelbick,[3,4] on the basis of structural features, the difference between them is substantial. Depending on the scale, an intimate mixture of the same components may exhibit different properties. Regardless of the scope of the

Food Chemistry, Function and Analysis
Bio- and Nano-sensing Technologies for Food Processing and Packaging
Edited by Ashutosh Kumar Shukla

Published by the Royal Society of Chemistry, www.rsc.org

definition, the important attribute of hybrid materials is a synergetic combination of inorganic and organic components while they are not just simple physical mixtures of the constituents. Therefore, the properties of these materials are not the sum of the individual contributions of both phases but are dependent on the inner interfaces between them. Taking into account the nature of the interface, hybrid materials can be divided into two classes. In class I, the chemical interactions between organic and inorganic components are weak (hydrogen, van der Waals or electrostatic interactions) resulting in blends or interpenetrating networks. In class II, the two phases are linked together through strong chemical bonds (covalent, coordinative bonds).[4,5]

Hybrid materials combine desirable features of organic polymers and inorganic components and thus offer several advantages. They exhibit toughness, elasticity and formability as well as hardness, strength and chemical resistance,[5,6] showing properties superior to those of their unmodified counterparts. Another important asset is that they can be processed as gels, monoliths, thin films, fibres, particles or powders.[7,8] Depending on the synthetic strategy, the performance of essential characteristics of the hybrids' properties can be tailored, resulting in multifunctional materials suitable for several applications. The variety of organic–inorganic hybrids seems to be unlimited and many have already become commercial products.[9] The possibility to design their features allows their usage in a wide range of industry sectors: textile, packaging, construction, automobile, micro-optics, microelectronics, as well as in the synthesis of functional coatings or biosensors, biocatalysts, and novel materials for dentistry, cosmetics, therapeutic vectors or biomedical purposes.[4,10–12]

4.1.1 Methods for the Synthesis of Hybrid Materials

Current developments in the synthesis of hybrids have resulted in numerous examples of materials that exhibit unique properties and encompass a wide variety of applications. However, it is worth noting that many of them are patterned on those existing in nature since innovative solutions are often inspired by the biomimetic design. Among the natural materials consisting of organic and inorganic building blocks are crustacean carapaces, mollusc shells and bone or teeth tissues in vertebrates.[13] Such composites combine inorganic building blocks which provide mechanical strength with organic parts which serve as a bonding.[4] One of the oldest representative and remarkable examples of class I hybrid material invented by humankind is Maya blue pigment. It is unique not only due to its beautiful colour but also its durability as it has withstood more than 12 centuries of a harsh jungle environment. Maya blue was prepared with molecules of the natural indigo encapsulated within the channels of a clay mineral known as palygorskite. Therefore, it combines the colour of organic pigment and the resistance of the inorganic host. Since ancient times, other examples of hybrid paints have been manufactured which consisted of inorganic nano-pigments suspended in organic solvents. The successful commercialization of organic–inorganic hybrid materials started in the 1950s. However, the concept of modern hybrid materials gained importance in the 1980s with the expansion of soft inorganic chemistry processes and development of the sol–gel technique.[5,14–18]

There are two main strategies to the formation of hybrid materials: *in situ* formation of the components and the building blocks approach. The *in situ* strategy involves chemical reaction of the precursors which are transformed to multidimensional structures. In the building block approach, well-defined preformed building units are applied in order to form the hybrid. Their structure (molecular integrity), at least partially, remains unchanged during the process of matrix formation, thus their structural units can be found in the final material. Moreover, building blocks also retain their typical properties.[4,19,20]

4.1.1.1 *Sol–Gel Route for* In Situ *Formation of Hybrid Materials*

Sol–gel synthesis offers several advantages such as mild synthetic conditions, low processing temperatures and the versatility of the forming procedures which can be applied for the preparation of the final hybrid material. In the sol–gel route, a liquid mixture of components is transformed into a three-dimensional cross-linked network which is formed as a result of the hydrolysis and condensation reactions that take place subsequently. Within the typical process, the transition from sol to gel takes place followed by the evaporation of a solvent. The possibility and easiness of control of the conditions of the sol–gel reaction enable materials to be obtained with controlled structures and properties. Figure 4.1 shows a schematic representation of the sol–gel route presenting a variety of final products: films, fibres, powders, patterns and monoliths. General strategies for the synthesis of sol–gel-derived hybrid materials have been extensively discussed in several reviews.[4,21–23]

In a typical silicon-based sol–gel process, R_{4-n}–SiX_{n-1} precursors are commonly used, where X represents OR′ or halogen ($n = 1$–4). The Si–X bond is labile towards hydrolysis reactions forming unstable silanol (Si–OH) that undergoes condensation reactions resulting in Si–O–Si bonds. A sol consists

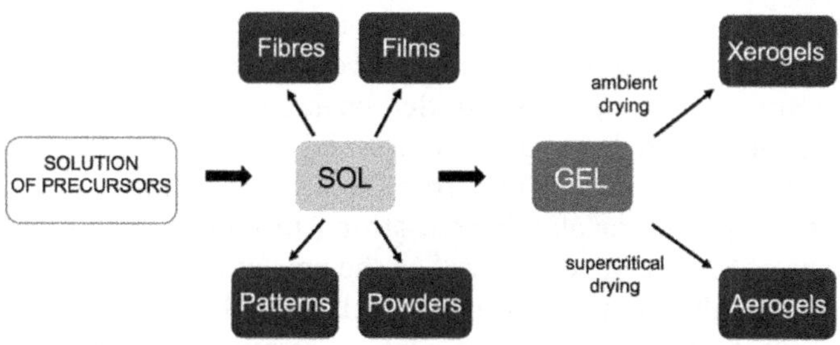

Figure 4.1 Sol–gel route.[4,21–23]

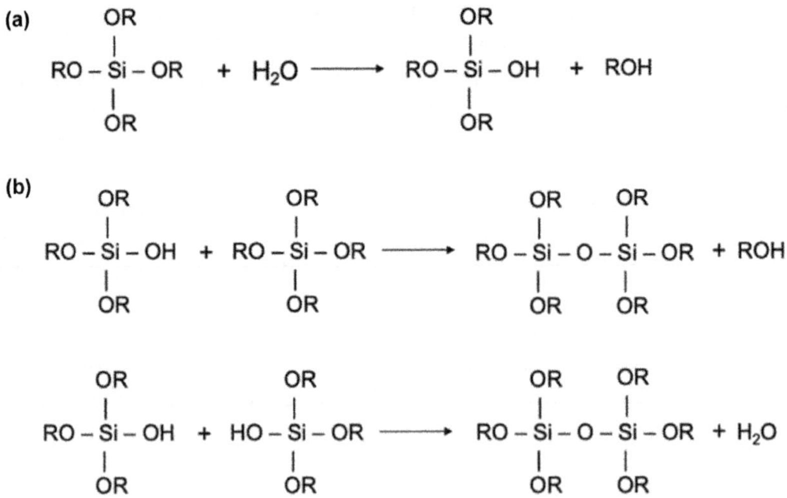

Figure 4.2 Chemical reactions of the sol–gel process of alkoxysilanes: (a) hydrolysis, (b) condensation.[4,23]

of colloidal particles (oligomers, polymers and cyclics) with diameters in the range of 1 ± 1000 nm which are dispersed in a liquid. In the sol–gel processing of alkoxysilanes, condensation takes place even if not all of the OR groups are hydrolysed. As intermediate species still contain Si–OR and/or Si–OH groups, hydrolysis occurs parallel to condensation reactions (Figure 4.2). A gel, formed during cross-linking reactions of solid particles present in a sol, consists of a spongelike, three-dimensional solid network whose pores are filled with another substance, usually a liquid (water and/or alcohol).[4,23] Depending on the drying procedure, supercritical or at ambient conditions, aerogels or xerogels are formed.[23]

The sol–gel process is determined by several parameters: type and concentration of the precursor, the relative concentration of the components in the precursor mixtures, the kind of solvent, the ratio of the water to alkoxy groups, the temperature and the pH value. Depending on the type of catalyst used for the sol–gel reaction, different mechanisms take place affecting the kinetics and resulting in the varied microstructure of the final material. Under an acid-catalysed process hydrolysis is favoured, leading to the formation of a great number of monomers or oligomers with reactive Si–OH groups. As a result, a polymerlike open network is formed (containing chains with few branches). In contrast, under basic conditions, a network consisting of particles and large pores is formed. As catalysts, HCl, NaOH or NH_4OH are commonly used. Apart from reaction conditions, the kind and structure of the precursor also have a strong influence on the sol–gel process. The size of the precursor, as well as the type of its substituents, have an impact on the solubility in the solvent as well as on the reaction time. Typical silicon precursors are alkoxysilanes such as tetramethoxysilane (TMOS) or tetraethoxysilane (TEOS), however in this case the compounds obtained during the sol–gel process are pure inorganic materials. Therefore, in order to obtain organic–inorganic hybrid materials, alkoxysilanes with the chemical formula $(RO)_{3-x}SiR_{1+x}$ can be applied as co-precursors. They provide additional groups that can undergo further chemical reactions and are responsible for new functions of the resulting material. As already mentioned, control of the reaction conditions is crucial in order to obtain hybrids with well-defined structure and desired features. Sol–gel processing enables the formation of organic–inorganic hybrids characterized with tailored mechanical, thermal, optical or physicochemical properties suitable for numerous applications.[3,4,9,23–25]

Due to the mild temperatures of the sol–gel process, organic components can be directly added to the sol and physically entrapped within the inorganic network formed during gelation. This type of hybrid material can be used in catalytic or sensor applications, while the diffusion of molecules to the entrapped compounds is possible through the porous sol–gel matrix. However, without a strong chemical interaction between two phases the stability of the material is limited and leaching may occur. Thus, chemical modification and application of organically functionalized trialkoxysilanes which provide covalent bonding between two phases are applied in order to obtain hybrids that do not show this disadvantage.[4] The sol–gel process can be performed not only in the presence of a small organic or biological structure (such as an enzyme or protein) but also in the presence of a polymer. The reaction takes place with or without chemical bonding between the two phases, depending on the functional groups of the compounds. Another approach for organic–inorganic hybrid preparation is simultaneous formation of interpenetrating networks (IPNs) of inorganic phase and polymer chain. In this procedure, sol–gel precursors are mixed with appropriate, selected monomers (vinyl ones) and both reactions are carried out at the same time. As a result, highly homogeneous materials are obtained. IPN hybrids also can be prepared by mixing the precursor for the sol–gel process with the organic polymer followed by an inorganic polycondensation reaction. Inorganic components can be incorporated into the polymer chain as molecular building blocks. The amounts of incorporated inorganic moieties as well as polymer types are high, therefore there are numerous examples of organic–inorganic hybrids, representing materials from class I and II. The ability to design well-defined hybrid structures, also at the nanoscale, results in a wide range of materials exhibiting interesting properties and that are suitable for varied applications such as electronic, optical, magnetic, biomedical and packaging industry uses. Special attention can be paid to organic–inorganic hybrids based on biopolymers while such materials serve as environmentally friendly alternatives to conventional synthetic polymers.[3,4,26–32] Wojciechowska and Foltynowicz[26] developed class I hybrid materials based on cellulose acetate butyrate (CAB) using different amounts of tetraethoxysilane (TEOS) as the sol–gel precursor. FT-IR analysis confirmed the formation of an inorganic phase as well as a ^{29}Si NMR spectrum that revealed the presence of Q^2, Q^3, Q^4 species related to the *in situ* formed silica network. SEM images of the samples obtained with different CAB/TEOS weight ratios (9.37/6.25, 87.5/12.5, 75/25 and 50/50) showed uniform distribution inorganic phase (disc-shaped) within the polymer matrix (Figure 4.3). CAB hybrids exhibited good transparency and enhanced oxygen transmission rate (OTR)[33] in comparison with a neat biopolymer showing potential for packaging applications.

In another work, Wojciechowska, Foltynowicz and Nowicki[28] described the preparation of hybrid materials based on modified cellulose acetate propionate (MCAP) and a covalently bonded inorganic phase. Prior to the sol–gel process, CAP was modified in the reaction of a 3-isocyanatopropyltriethoxysilane (ICPTS) coupling agent. Next, the inorganic phase was *in situ* introduced into the MCAP matrix using TEOS in MCAP/TEOS weight ratios of: 93.75/6.25 and 87.5/12.5. The structure was confirmed using FT-IR, solid-state ^{29}Si NMR, as well as SEM images. Moreover, AFM analysis revealed a correlation between the amount of applied TEOS and the inorganic-rich region size, dispersion and RMS roughness of the obtained hybrids (Figure 4.4). Figure 4.4 shows AFM images revealing the size and distribution of the inorganic phase (white spots) in the dark polymer matrix. In another work, Wojciechowska and Kozak[33] confirmed that chemical modification of CAP leads to the successful improvement of barrier properties in comparison with samples based on an unmodified biopolymer. The longer diffusion pathway of oxygen molecules observed in MCAP samples was a consequence of the smaller dimensions of the inorganic phase introduced into the polymer matrix. MCAP organic–inorganic hybrids show potential for

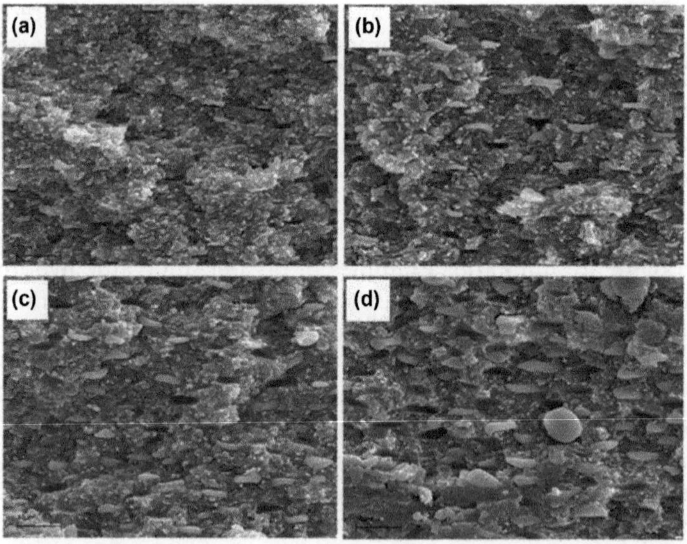

Figure 4.3 SEM images of organic–inorganic hybrids prepared with different CAB/TEOS weight ratios: (a) CAB6.25, (b) CAB12.5, (c) CAB25, (d) CAB50. Reproduced from ref. 26 with permission from Industrial Chemistry Institute, Copyright 2009.

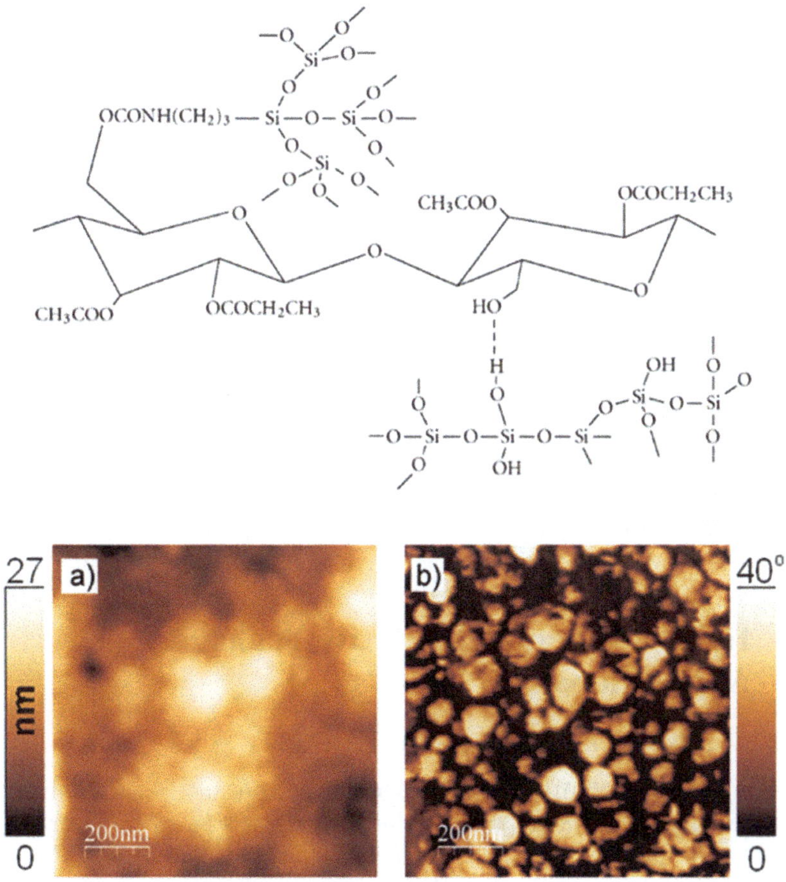

Figure 4.4 Chemical structure (top) and AFM images of MCAP12.5 hybrid (bottom): (a) topography, (b) material contrast. Reproduced from ref. 28, DOI: 10.1155/2013/616159, under the terms of the CC BY 3.0 license, https://creativecommons.org/licenses/by/3.0/.

packaging applications as an ecologically friendly alternative to conventional petrochemical polymers.

4.1.1.2 Building Block Approach

Another common route for the synthesis of hybrid materials is a building block approach. Organic, as well as inorganic structures can be incorporated into a second phase, however compatibility between the components is crucial. Building blocks should be characterized by defined structure, size and shape and ability to introduce an interesting feature to the final material. Typical inorganic building blocks with well-defined structures are cluster compounds, usually surface-functionalized in order to improve compatibility with an organic phase. Some of the examples are spherosilicates, polyhedral oligomeric silsesquioxanes (POSS) or metal oxo clusters with attached reactive organic groups. The application of nanoparticles or nanorods results in the formation of nanocomposites. Among organic building blocks, oligomers, polymers and biological active molecules are applied. The main advantage of the building block approach is easy control of the final material properties due to the fact that well-defined building blocks keep their molecular integrity, as well as typical features, during the hybrid formation.[4]

An interesting strategy for hybrid materials formation from well-defined nanobuilding blocks (NBB), *via* assembling or intercalation or intercalation and dispersion, is described and discussed by Sanchez and co-workers. As NBBs, clusters, organically pre- or post-functionalized nanoparticles (metallic oxides, metals, chalcogenides, *etc.*), nano-core–shells or layered compounds (clays, layered double hydroxide, lamellar phosphates, oxides or chalcogenides) can be applied. Among organic–inorganic hybrids using NBBs, materials based on layered silicates incorporated into the polymer matrix show noticeable commercial potential, especially in construction, automotive, sealant, textile, wood as well as packaging applications.[21]

4.2 Hybrid Materials for Active Packaging Applications

4.2.1 Active Packaging Systems

According to the definition provided by the European Commission Regulation No. 450/2009[34] *active materials and articles are intended to extend the shelf-life or to maintain or improve the condition of packaged food.* Active materials and articles may deliberately incorporate substances which are intended to be released or absorbed into or from the packaged food or the environment surrounding the food (Figure 4.5).

In other words, in an active packaging system, there is an intentional interaction between the packed product, packaging and the environment.

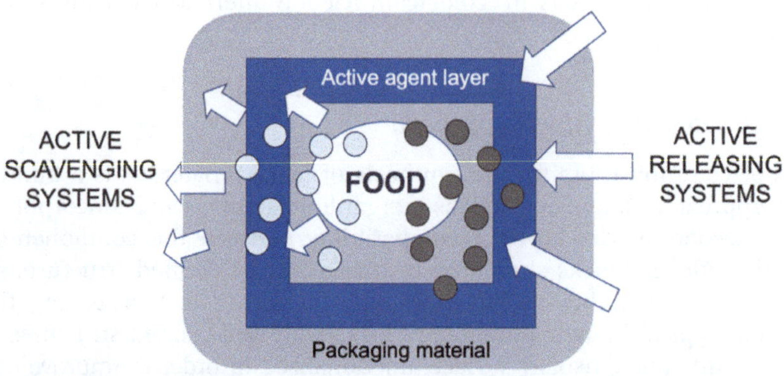

Figure 4.5 Active packaging systems.

In contrast, conventional food packaging works as a passive barrier protecting food against environmental contamination. A significant advantage of an active systems application in food packaging is quality maintenance and product preservation resulting in an extended shelf life of packaged food. This can be achieved mainly due to the desired impact of active systems on varied processes specific for each type of food: breathing of fresh fruit and vegetables, moisture loss of bakery products, oxidation of fats or myoglobin in raw meat, maturating of cheese, microbiological changes, *etc.*[35]

Active substances can be placed inside the package (as a pad, blanket) or in a separate container (sachet), or can be incorporated directly into the packaging material. There are two main groups of these systems: scavengers (absorbers) and emitters (releasers). Scavengers remove undesired compounds from the packaging, improving the conditions for storage and thus extend the shelf life of the food. The most well-known moisture absorbers are based on zeolite, cellulose or their derivatives. They are mostly introduced into trays for raw meat or fish. In addition to moisture regulators, oxygen, carbon dioxide, flavour/odour absorbers and ethylene scavengers are applied. Other substances applied as absorbers include: activated carbon, silica gel, iron ions, ascorbic acid, potassium permanganate and calcium hydroxide. Among active releasing systems, antimicrobial agents, carbon dioxide, antioxidant compounds, ethylene, flavourings or preservatives are applied. Emitters release desired compounds that change the environment inside the package exhibiting a positive impact on the product.[35–37] Some examples of commercially available systems are[36] Activ-Film™ (moisture absorber), BIPAC (ethylene scavenger), ATCO® (carbon dioxide absorber), FreshPax® (carbon dioxide emitter), Celox™ (oxygen scavenger) and Biomaster® (antimicrobial).

Antimicrobial packaging systems release active compounds in order to delay the growth of undesired microbes. They restrain microbiological deterioration which extends the shelf life of the packaged food and provides consumers with protection against foodborne pathogens.[38] Common active agents applied in packaging for food products are organic acids (propionic, acetic, sorbic or benzoic acids) and their salts (potassium or calcium sorbate, sodium benzoate), bacteriocins, enzymes (nisin, lysozyme), alcohols (ethanol), sulphites, nitrites, silver nanoparticles and different natural extracts and essential oils (thymol, oregano or eucalyptus essential oils). According to the safety issues, all commercial antimicrobial agents require food contact approval.[39–44] There are varied active systems that can be applied in food packaging: sachets or mats releasing volatile compounds, polymer films with incorporated active agents or compounds immobilized on their surface, and biopolymers revealing antimicrobial activity (chitosan).[35] The effectiveness of a chosen antimicrobial agent, regarding its application in active packaging systems, depends on several factors. The residual antimicrobial activity of a particular packaging material is influenced by the chemical nature, polarity and compatibility of the polymer. Moreover, heat lability of biologically active agents such as lysozyme should be taken into consideration when it comes to common processing techniques in the packaging industry.[45]

4.2.2 Gelatine–Siloxane Hybrids for Active Packaging Systems

In recent years, active films based on biopolymers have received significant attention. This is due to the growing environmental concern and demand for alternatives of conventional synthetic polymers. However, many natural polymers exhibit low resistance against external factors such as high humidity, elevated temperatures as well as susceptibility to the impact of packaged food. Therefore, synthetic strategies for organic–inorganic hybrid preparation enable the modification of biopolymers in order to obtain materials with enhanced properties, suitable for biological or packaging applications.[9,32]

Wojciechowska *et al.*[45] developed antimicrobial films based on biodegradable and biocompatible gelatine using antimicrobial agents such as lysozyme, ethylenediaminetetraacetic acid disodium dihydrate (EDTA) and nisin. Gelatine was chemically modified with organofunctionalized silicones: 3-(3-glycidoxypropyl)heptamethyltrisiloxane (ES) and poly[dimethyl][methyl,3-(2,2,3,3,4,4,5,5-octafluoropentyloxy)propyl]siloxane (EFS). Active films based on gelatine–siloxane hybrids were prepared by a solvent casting method. The antimicrobial activity of the active hybrids was examined using a modified agar diffusion method against selected bacterial strains (*Aeromonas hydrophila*, *Bacillus subtilis*, *Escherichia coli*) and opportunistic fungus (yeast species *Candida albicans*). Films based on neat gelatine and gelatine–siloxane hybrid were also prepared and analysed as a reference. The results showed that gelatine–siloxane hybrids revealed enhanced mechanical properties in comparison with unmodified, brittle gelatine. Moreover, active films obtained with EFS siloxane exhibited higher effectiveness of introduced antimicrobial compounds than hybrids prepared with ES and films based on neat biopolymer. It was also confirmed that adjustment of the type of active agent to the target microorganism is crucial for antimicrobial activity of the films (EDTA was effective for *C. albicans* fungi, lysozyme for *B. subtilis* bacteria). Gelatine–siloxane hybrid films showed an insignificant change in colour and appearance in comparison with unmodified gelatine film. They also exhibited lower water uptake and better water stability than neat biopolymer, revealing their potential for packaging applications. In another work, Wojciechowska and Kozak[46] demonstrated that organic–inorganic hybrids based on gelatine revealed enhanced barrier properties comparing to unmodified biopolymer (Table 4.1). Depending on the type of organomodified siloxane used for the synthesis water vapour transmission rate (WVTR), as well as oxygen transmission rate (OTR), were shown to have up to three times lower values in comparison with neat gelatine.

Wojciechowska and Tichoniuk also confirmed the antimicrobial activity of gelatine–siloxane hybrid films prepared with organomodified siloxane and grapefruit oil as an active agent against common plant pathogens: *Erwinia carotovora* and *Thielaviopsis basicola*. Figure 4.6 presents the inhibition zones observed in the case of active hybrid films (H) in comparison with unmodified gelatine (G).

Table 4.1 WVTR and OTR values of gelatine–siloxane hybrids and unmodified biopolymer. Adapted from ref. 46 with permission from Wydawnictwo Uniwersytetu Ekonomicznego w Poznaniu, Copyright 2018.

Type of material	WVTR (g m^{-2} 24 h)	OTR (cm^3 m^{-2} 24 h)
Gelatine	100 ± 2	2.1 ± 0.06
Gelatine–siloxane hybrid 1	34 ± 1	1.0 ± 0.05
Gelatine–siloxane hybrid 2	53 ± 1.5	0.7 ± 0.05
Gelatine–siloxane hybrid 3	65 ± 2	—

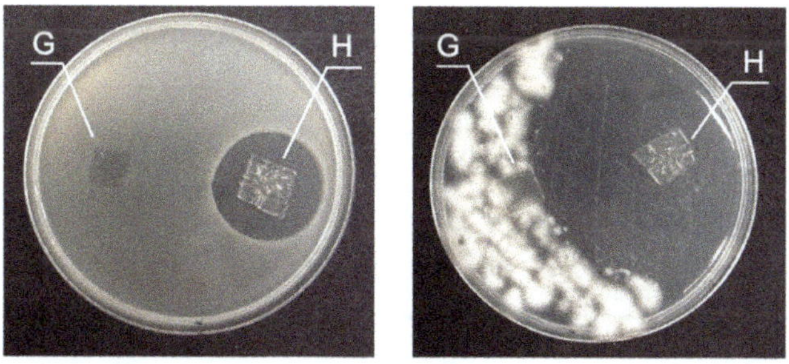

Figure 4.6 Antimicrobial activity of gelatine–siloxane hybrid films (H) against *Erwinia carotovora* (left) and *Thielaviopsis basicola* (right) in comparison with gelatine (G).

In another work, Tichoniuk, Wojciechowska and Maciejewski[43] examined the antimicrobial activity of gelatine–siloxane hybrid films against *Erwinia carotovora* and *Thielaviopsis basicola* using oregano and eucalyptus essential oils, as well as the monoterpene phenols, thymol and carvacrol. The results showed that the most effective active agents for the examined active films based on unmodified gelatine were thymol and carvacrol. In contrast, all gelatine–siloxane hybrid active films exhibited antimicrobial properties against *Erwinia carotovora* and *Thielaviopsis basicola*. The largest inhibition zones were observed for samples enriched with oregano and carvacrol (see Table 4.2). The obtained results confirmed that organic–inorganic hybrids show potential for packaging applications and may enhance consumer safety, as well as the shelf life of unprocessed food products of plant origin such as raw fruit and vegetables.

Gelatine–siloxane hybrids can also be applied as a coating on paper or cardboard in order to enhance the functional parameters of this type of packaging material. Wojciechowska[46] demonstrated that they provide a protective layer that enhanced the barrier properties of a paper substrate. In the case of paper covered with a hybrid layer with a thickness of 27.3 μm the OTR value was reduced to 3.1 cm^3 m^{-2} 24 h (Figure 4.7). Moreover, it was shown that modification of its surface with an SiOx coating using chemical vapour deposition (CVD) resulted in greater resistance to moisture (Figure 4.8).

Table 4.2 Antimicrobial activity of gelatine and gelatine–siloxane hybrid films against *Erwinia carotovora* and *Thielaviopsis basicola*. Reproduced from ref. 43 with permission from the *Polish Journal of Commodity Science*, Copyright 2021.

Gelatine (G) or gelatine–siloxane hybrid (H) film	Active agent	*E. carotovora*	*T. basicola*
		Inhibition zone (mm)	
G	None	—	—
GT	Thymol	Contact-based limited growth	Contact-based limited growth
GE	Eucalyptus	—	Contact-based limited growth
GO	Oregano	—	Contact-based inhibition
GC	Carvacrol	Contact-based inhibition	Contact-based inhibition
H	None	30.8	38.0
HT	Thymol	33.6	39.0
HE	Eucalyptus	31.6	>40.0
HO	Oregano	35.4	>40.0
HC	Carvacrol	35.4	>40.0

Figure 4.7 SEM image of a gelatine–siloxane hybrid coating applied on a paper substrate. Reproduced from ref. 46 with permission from Wydawnictwo Uniwersytetu Ekonomicznego w Poznaniu, Copyright 2018.

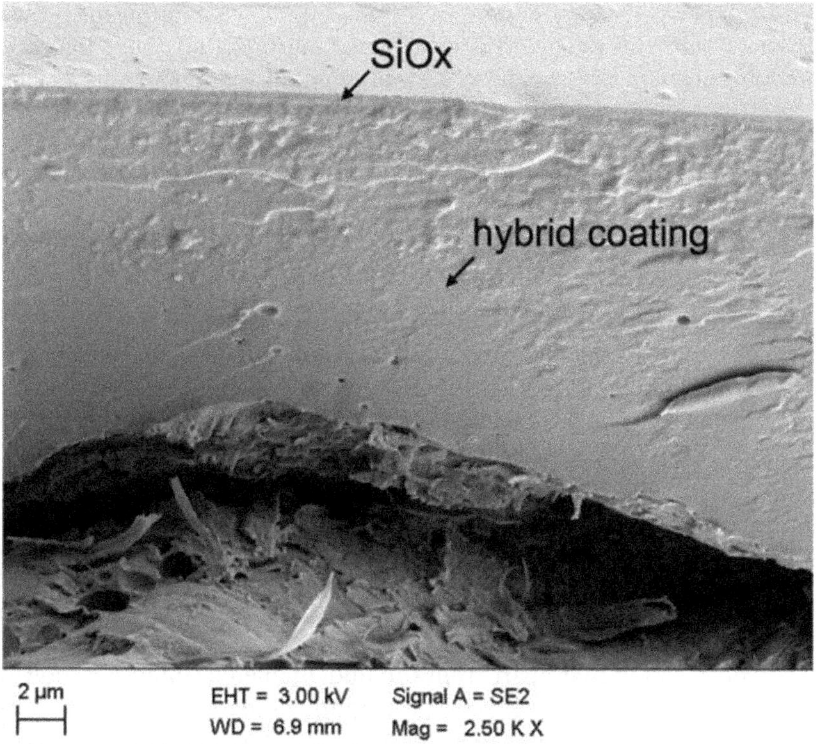

Figure 4.8 SEM image of a gelatine–siloxane hybrid coating applied on a paper substrate and covered with a SiOx layer. Reproduced from ref. 46 with permission from Wydawnictwo Uniwersytetu Ekonomicznego w Poznaniu, Copyright 2018.

Table 4.3 Physical parameters of paper substrates covered with an organic–inorganic hybrid layer. Reproduced from ref. 46 with permission from Wydawnictwo Uniwersytetu Ekonomicznego w Poznaniu, Copyright 2018.

Sample	Thickness (µm)	Grammage ($g\,m^{-2}$)	Tensile strength ($kN\,m^{-1}$)	Elongation at break (%)	Bursting strength (kPa)
P0	113 ± 1	82.9 ± 0.4	1.9 ± 0.2	2.2 ± 0.3	70.4 ± 2.1
H1–50	125 ± 1	90.5 ± 0.6	2.4 ± 0.1	3.3 ± 0.1	67.8 ± 2.6
H1–70	134 ± 1	96.1 ± 0.9	3.0 ± 0.1	3.8 ± 0.3	69.3 ± 3.0

Further investigation revealed that organic–inorganic hybrid coatings enhance the mechanical properties of paper substrates, which is an important feature in the case of packaging materials (Table 4.3).

The data presented in Table 4.3 show that a paper substrate covered with a layer of organic–inorganic hybrid exhibited enhanced mechanical properties (tensile strength) and higher elongation at break. In this case, the impact on bursting strength was not evident. The results confirmed that hybrid materials, also enriched with antimicrobial agents, can serve as a functional layer for paper packaging materials. Taking into consideration the growing interest of the packaging industry in the application of paper instead of synthetic polymers, this solution shows potential for fresh fruits (such as raspberries, blueberries or strawberries) or vegetables (carrots, parsley and celery) packaging.

4.2.3 Organic–Inorganic Hybrid Nanocomposites for Active Packaging

There are numerous reports in the scientific literature devoted to organic–inorganic hybrid nanocomposites, based on conventional fossil-based polymers as well as biopolymers, enhanced with inorganic compounds.[47–51] A wide group of active materials is composed of nanosized compounds introduced into the polymer matrix, such as metal oxide nanoparticles (*e.g.* silver, titanium, copper, zinc, ZnO, MgO and TiO_2) or layered silicates (montmorillonite) combined with antimicrobial agents.[36,48] The suggested mechanism of antimicrobial action based on the release of metal ions is the formation of reactive oxygen species which are responsible for damage to the walls and membranes of microbial cells.[36] Some examples of active polymer nanocomposites are listed in Table 4.4.

It is worth noting that apart from the undeniable advantages of polymer nanocomposites, the benefits for packaging industry such as enhanced mechanical, barrier or chemical properties, utilization of nanocomposites in food contact materials (FCMs) and societal acceptance of nanotechnology remain open issues. The possible migration of nanocomponents from FCMs is one of the most important concerns in human exposure due to the

Table 4.4 Examples of antimicrobial polymer nanocomposites.[52–61]

Active agent	Polymer	Antimicrobial activity	References
Ag nanoparticles	Polylactide–polyethyleneimine	*S. aureus*, *E. coli*	52
Ag nanoparticles	Chitosan–starch	*S. aureus*, *E. coli*	53
Ag nanoparticles	Polyvinyl alcohol	*Salmonella typhimurium*	54
TiO_2	Low-density polyethylene	*Rhodotorula mucilaginosa*, *Pseudomonas* spp.	55
TiO_2–Ag	Gelatine–chitosan	*E. coli*	56
ZnO	Gelatine–tragacanth	*S. aureus*, *E. coli*	57
ZnO	Regenerated cellulose	*B. cereus*, *S. aureus*, *L. monocytogenes*, *E. coli*, *S. typhimurium*, *V. parahaemolyticus*	58
ZnO–SiO_2	Polyvinyl alcohol–chitosan	*S. aureus*, *E. coli*	59
Carvacrol	Methyl cellulose/montmorillonite	—	60
Carvacrol	Thermoplastic starch/montmorillonite	*E. coli*	61

potential health risks. Therefore, risk assessment of nanomaterials designed for food packaging materials plays a crucial role in food safety. In the European Union, Regulation (EU) No. 10/2011 on plastic materials and articles intended to come into contact with food[62] sets out rules on the composition of plastic FCMs. This document establishes a Union List of substances that are permitted for use in the manufacture of plastic FCMs. It also specifies restrictions on the use of some substances and sets out rules to determine the compliance of plastic materials and articles.[62] According to this regulation, the overall migration limit is 10 $mg\,dm^{-2}$ (60 $mg\,kg^{-1}$ food for cubic packaging with 1 kg of food). Regulation (EU) No. 10/2011 specifies that the assessment of nanoparticles should be undertaken each time on a case-by-case basis, under standardized test conditions: testing time, temperature and test medium (food simulant). Annex I of this regulation authorizes only three types of nanoparticles, silica, carbon black and titanium nitride, which can be used in plastic packaging for food (including nanoparticles that are intended to be used behind a functional barrier). In the case of plastic FCMs containing particles or fillers other than the three listed above, an approval of the European Food Safety Authority (EFSA), based on specific information regarding the migration, toxicology and possible exposure is mandatory for their authorization.[48]

Organic–inorganic hybrids are multifunctional materials showing great potential in numerous applications, including: optics, electronics, catalysis, sensors, environmental remediation, cosmetics, textiles, construction, automobiles, as well as in the biomedical and packaging industries. They can be successfully combined with active agents resulting in antimicrobial packaging materials which show desirable mechanical properties, permeability of gases and formability. They also offer quality maintenance and extended shelf life of the packaged food, as well as enhanced consumer safety.

References

1. IUPAC, Compendium of Chemical Terminology, (the "Gold Book"), compiled by A. D. McNaught and A. Wilkinson, Blackwell Scientific Publications, Oxford, 1997, 2nd edn, online version (2019) created by S. J. Chalk, ISBN 0-9678550-9-8, DOI: 10.1351/goldbook.
2. Commission Recommendation of 18 October 2011 on the definition of nanomaterial (Text with EEA relevance) (2011/696/EU).
3. G. Kickelbick, *Hybrid Mater.*, 2014, **1**, 39.
4. G. Kickelbick, *Hybrid Materials: Synthesis, Characterization, and Applications*, Wiley-VCH Verlag GmbH & Co. KGaA, Weinheim, 2007.
5. C. Sanchez, B. Julian, P. Belleville and M. Popall, *J. Mater. Chem.*, 2005, **15**, 3559.
6. S. Pandey and S. B. Mishra, *J. Sol–Gel Sci. Technol.*, 2011, **59**, 73.
7. G. Kickelbick, *Prog. Polym. Sci.*, 2003, **28**, 83.
8. S. Gross and M. Bauer, *Adv. Funct. Mater.*, 2010, **20**, 4026.

9. P. Wojciechowska, P. Pietras and H. Maciejewski, *Adv. Polym. Technol.*, 2014, **33**(S1), DOI: 10.1002/adv.21459.
10. G. L. Drisko and C. Sanchez, *Eur. J. Inorg. Chem.*, 2012, **32**, 5097.
11. K. Tsuru, S. Hayakawa and A. Osaka, *J. Sol-Gel Sci. Technol.*, 2004, **32**, 201.
12. S. Wang, Y. Kang, L. Wang, H. Zhang, Y. Wang and Y. Wang, *Sens. Actuators, B*, 2013, **182**, 467.
13. C. Sanchez, H. Arribart and M. Giraud Guille, *Nat. Mater.*, 2005, **4**, 277.
14. B. Arkles, *CHEMTECH*, 1999, **29**, 7.
15. Y. Chujo, *Curr. Opin. Solid State Mater. Sci.*, 1996, **1**, 806–811.
16. C. Sanchez and F. Ribot, *New J. Chem.*, 1994, **18**, 1007.
17. P. Judeinstein and C. Sanchez, *J. Mater. Chem.*, 1996, **6**, 511.
18. U. Schubert, N. Hüsing and A. Lorenz, *Chem. Mater.*, 1995, 7(11), 2010.
19. S. K. Pillali and S. S. Ray, Inorganic-Organic Hybrid Polymers for Food Packaging, in *Functional Polymers in Food Science. From Technology to Biology, Food Packaging*, ed. G. Cirillo, U. G. Spirizzi and F. Iemma, Scrivener Publishing LLC, Salem, Massachusetts, 2015, vol. 1.
20. S. Pandey and S. B. Mishra, *J. Sol–Gel Sci. Technol.*, 2011, **59**, 73.
21. C. Sanchez, P. Belleville, M. Popall and L. Nicole, *Chem. Soc. Rev.*, 2011, **40**(696), 753.
22. C. Sanchez, L. Rozes, F. Ribot, C. Laberty-Robert, D. Grosso, C. Sassoye, C. Boissiere and L. Nicole, *C. R. Chim.*, 2010, **13**(1–2), 3.
23. N. Hüsing and U. Schubert, *Angew. Chem., Int. Ed.*, 1998, **37**, 22.
24. J. Karasiewicz, A. Dutkiewicz and H. Maciejewski, *Chemik*, 2014, **68**(11), 945.
25. H. Maciejewski, I. Dąbek and B. Marciniec, *Polimery*, 2004, **49**(10), 677.
26. P. Wojciechowska and Z. Foltynowicz, *Polimery*, 2009, **54**(11–12), 845.
27. P. Wojciechowska, Z. Foltynowicz and M. Nowicki, Cellulose Acetate Butyrate Nanocomposites Synthesized via Sol-Gel Method, *Polimery*, 2013, **58**(7/8), 543.
28. P. Wojciechowska, Z. Foltynowicz and M. Nowicki, *J. Spectrosc.*, 2013, **616159**, 1.
29. T. Krasia-Christoforou, Organic-inorganic polymer hybrids: Synthetic strategies and applications, *Hybrid and Hierarchical Composite Materials*, Springer, Berlin/Heidelberg, Germany, 2015.
30. G. Kickelbick, *Prog. Polym. Sci.*, 2003, **28**, 83.
31. *Hybrid Nanocomposites for Nanotechnology. Electronic, Optical, Magnetic and Biomedical Applications*, ed. L. Merhari, Springer Science + Business Media, LLC, New York, 2009.
32. *Bio-inorganic Hybrid Nanomaterials. Strategies, Syntheses, Characterization and Application*, ed. E. Ruiz-Hitzky, K. Ariga and Y. Lvov, Wiley-VCH Verlag GmbH & Co. KGaA, Weinheim, 2008.
33. P. Wojciechowska and W. Kozak, Towaroznawcze Problemy Jakości, *Pol. J. Commod. Sci.*, 2016, **4**(49), 81.
34. European Commission Regulation No. 450/2009 of 29 May 2009 on active and intelligent materials and articles intended to come into contact with food.

35. J. Wyrwa and A. A. Barska, *Eur. Food Res. Technol.*, 2017, **243**, 1681.
36. E. Drago, R. Campardelli, M. Pettinato and P. Perego, *Foods*, 2020, **9**, 1628.
37. R. Chawla, S. Sivakumar and H. Kaur, *Carbohydr. Polym. Technol. Appl.*, 2021, **2**, 100024.
38. K. K. Kuorwel, M. J. Cran, K. Sonneveld, J. Miltz and S. W. Bigger, *J. Food Sci.*, 2011, **76**, R90.
39. J. H. Han, *Food Technology*, 2000, **54**, 56.
40. P. Suppakul, K. Sonneveld, S. W. Bigger and J. Miltz, *J. Food Sci.*, 2003, **68**, 408.
41. P. Kanmani and J. W. Rhim, *Food Chem.*, 2014, **148**, 162.
42. B. Malhotra, A. Keshwani and H. Kharkwai, *Front. Microbiol.*, 2015, **6**, 1.
43. M. Tichoniuk, P. Wojciechowska and H. Maciejewski, Towaroznawcze Problemy Jakości, *Pol. J. Commod. Sci.*, 2021, **1**(66), 21.
44. M. Flórez, E. Guerra-Rodríguez, P. Cazón and M. Vázquez, *Food Hydrocolloids*, 2022, **124**, 107328.
45. P. Wojciechowska, M. Tichoniuk, D. Gwiazdowska, H. Maciejewski and M. Nowicki, *Adv. Polym. Technol.*, 2018, **37**, 2958.
46. P. Wojciechowska, *Materiały hybrydowe w innowacjach opakowaniowych (Hybrid materials for packaging applications)*, Poznań, 2018.
47. M. Tavassoli, M. A. Sani, A. Khezerlou, A. Ehsani and D. J. McClements, *Food Hydrocolloids*, 2021, **118**, 106747.
48. T.-G. Sarfraz, J. Nilsen-Nygaard and M. Kvalvåg Pettersen, *Nanomaterials*, 2021, **11**, 10.
49. M. Qian, D. Liu, X. Zhang, Z. Yin, B. B. Ismail, X. Yea and M. Guo, *Trends Food Sci. Technol.*, 2021, **114**, 459.
50. M. V. Dias, V. M. de Azevedo, L. F. Ferreira, A. C. S. de Oliveira, S. V. Borges, N. de Fátima Ferreira Soares, E. A. A. Medeiros and J. de Deus Souza Carneiro, *J. Food Process Eng.*, 2021, **44**(11), e13843.
51. S. Y. H. Abdalkarim, L.-M. Chen, H.-Y. Yu, F. Li, X. Chen, Y. Zhou and K. C. Tam, *Int. J. Biol. Macromol.*, 2021, **182**, 1915.
52. V. Demchenko, S. Kobylinskyi, M. Iurzhenko, S. Riabov, A. Vashchuk, N. Rybalchenko, S. Zahorodnia, K. Naumenko, O. Demchenko, G. Adamuse and M. Kowalczuk, *React. Funct. Polym.*, 2020, **170**, 105096.
53. J. Jung, G. Kasi and J. Seo, *Int. J. Biol. Macromol.*, 2018, **112**, 530.
54. R. Tripathi, R. N. Pudake, B. Shrivastav and A. Shrivastav, *Adv. Nat. Sci.: Nanosci. Nanotechnol.*, 2018, **9**, 025020.
55. H. Bodaghi, Y. Mostofi, A. Oromiehie, Z. Zamani, B. Ghanbarzadeh, C. Costa, A. Conte and M. A. Del Nobile, *LWT–Food Sci. Technol.*, 2013, **50**, 702.
56. D. Lin, Y. Yang, J. Wang, W. Yan, Z. Wu and H. Chen, *Int. J. Biol. Macromol.*, 2020, **154**, 123.
57. R. Shahvalizadeh, R. Ahmadi, I. Davandeh, A. Pezeshki, S. A. S. Moslemi, S. Karimi, M. Rahimi, H. Hamishehkar and M. Mohammadi, *Food Chem.*, 2021, **354**, 129492.

58. S. Saedi, M. Shokri, J. T. Kim and G. H. Shin, *J. Food Eng.*, 2021, **307**, 110665.
59. N. A. Al-Tayyara, A. M. Youssef and R. R. Al-Hindi, *Food Packag. Shelf Life*, 2020, **25**, 100523.
60. S. Tunç, O. Duman and T. G. Polat, *Carbohydr. Polym.*, 2016, **150**, 259.
61. A. G. de Souza, N. M. A. dos Santos, R. F. da Silva Torin and D. dos Santos Rosa, *Int. J. Biol. Macromol.*, 2020, **164**, 1737.
62. Commission Regulation (EU) No. 10/2011 of 14 January 2011 on plastic materials and articles intended to come into contact with food.

CHAPTER 5

Bio-based and Nanostructured Hybrids for Green and Active Food Packaging

MARIA-NEFELI EFTHYMIOU,[a] ERMINTA TSOUKO,[a] ELENI VLASSI,[b] ARISTEIDIS PAPAGIANNOPOULOS,*[b] APOSTOLIS KOUTINAS[a] AND STERGIOS PISPAS[b]

[a] Department Food Science and Human Nutrition, Agricultural University of Athens, IeraOdos 75, 11855 Athens, Greece; [b] Theoretical and Physical Chemistry Institute, National Hellenic Research Foundation, 48 Vassileos Constantinou Ave., 11635, Athens, Greece
*Email: apapagiannopoulos@eie.gr

5.1 Introduction

The large expansion of conventional and synthetic plastics used as food packaging materials has created several environmental issues such as disposal shortcomings, the greenhouse gas effect and toxic emissions. It has been reported that, in 2018, 113 billion food packages were consumed within the countries of the European Union. So far, 90% of materials that have been employed in plastic packaging applications include non-biodegradable and partially recyclable compounds, *i.e.* polyethylene, polyvinyl chloride and polypropylene.[1] Natural hybrid nanostructured materials have been an emerging field of investigation during the last decade due to their excellent properties of biocompatibility, biodegradability, thermal stability and mechanical response. They have been so far applied in the

Food Chemistry, Function and Analysis
Bio- and Nano-sensing Technologies for Food Processing and Packaging
Edited by Ashutosh Kumar Shukla

Published by the Royal Society of Chemistry, www.rsc.org

emerging sectors of catalysis, pharmaceuticals, medicine, the polymer industry and food packaging.[2] Proteins and polysaccharides isolated from agro-industrial by-product streams or microbial biopolymers produced by several microbial species could be ideal building blocks for alternative food packaging materials, while nanostructures of these could enhance the physical and mechanical properties of the final formulations.[3]

Biopolymers offer the versatility of macromolecular materials in terms of their ability to form soft and hierarchically organized nanostructures that can interact and incorporate other organic and inorganic molecules, and the biocompatibility and biodegradability of nature-derived polymers. From simple polysaccharides to complex proteins, the physicochemical properties of biopolymers such as charge, hydropathy and possible conformational transitions, *e.g.*, temperature-induced denaturation, can be used to design and develop novel biopolymer films. Some examples include the application of chitin and chitosan in bioplastic packaging,[4] functional biofilms of starch with advanced physicochemical properties[5] and antimicrobial films of carrageenan with metal oxide nanoparticle inclusions.[6] Advanced physicochemical characterization such as electron microscopy and small angle scattering are very effective in resolving the spatial arrangement of biopolymer films at the different length scales and assist in the optimization of the final product.[7]

This chapter covers the recent advances in biopolymer films for active food packaging, starting with a thorough presentation of the biopolymers in broad categories that include polysaccharides, proteins and lipid-based biopolymers, and polymers and biopolymers coming from bio-monomers and microbial biopolymers. The relationship between the physicochemical properties of the active films and their function is discussed on the basis of physicochemical characterization techniques with an emphasis on the nanostructure, and the progress on nanocellulose in particular. Active food packaging is analyzed on the basis of function, such as antioxidant properties, emission of CO_2, scavenging of O_2 and the inclusion of inorganic nanoparticles for the enhancement of active properties. This chapter may serve as an up-to-date guide for a detailed overview of active and green biopolymer films for food packaging applications.

5.2 Natural and Microbial-derived Biopolymers

Biopolymers are defined as natural polymeric compounds synthesized by plants, animals and microorganisms through complex enzymatic bioprocesses that lead to the formation of high-molecular-weight molecules.[8] Biopolymers are generally biodegradable and bio-based and they can serve as raw materials for bioplastics.[9] This chapter focuses on biopolymers derived mainly from renewable resources. Bio-based polymers are broadly categorized in naturally produced biopolymers from plant and animal sources, *i.e.* proteins, polysaccharides and lipids, artificially synthesized polymers derived from bio-based monomers *via* chemical synthesis and

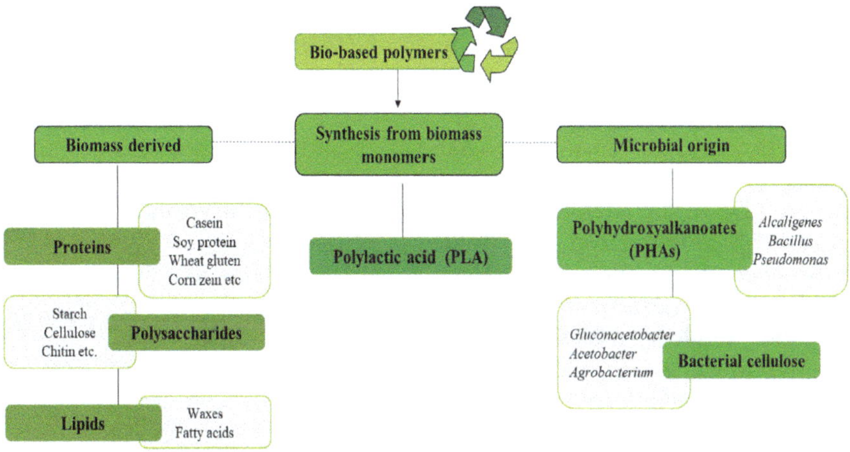

Figure 5.1 Origin and classification of bio-based polymers.

polymerization, such as poly(lactic acid) (PLA), and polymers produced intra- or extracellularly *via* microorganisms, including microbial polyesters [poly(hydroxyalkanoates); PHAs] and polysaccharides (bacterial cellulose; BC)[10] (see Figure 5.1).

Biopolymers derived from renewable resources present highly promising alternatives for sustainable, eco-friendly and non-toxic packaging materials and could effectively replace fossil-based plastics.[11,12] Film and coating forming ability of natural biopolymeric materials has gained great interest over the past decade due to the desirable structural, mechanical, water and oxygen barrier properties.[13–15]

5.2.1 Bio-based Polymers Derived from Renewable Resources

5.2.1.1 Polysaccharides

Polysaccharides are biodegradable macromolecules with linear or branched structures. Their functionality derives from different functional groups present in their molecule. Among polysaccharides, starch, cellulose (native or modified) and chitosan have been extensively studied in biodegradable packaging applications.[5,16–18] Their structural diversity confers distinct physicochemical properties[8] combined with great film-forming ability.[10] Conversion of polysaccharides into bio-based films and coatings, and their application to different edible products such as bread and fresh fruits have been reported.[15,19]

5.2.1.1.1 Cellulose. Cellulose is the most abundant polysaccharide in nature. Cellulose can be isolated from various natural sources such as wood, agricultural crops, grasses, agroforestry residues or industrial processed agricultural wastes, as well as tunicates and algae.[2] Cellulose is a linear homopolysaccharide consisting of β-D-glycose units linked *via* β-(1,4)-glycosidic bonds.[3] Native cellulose is highly crystalline, insoluble in water and presents a poor mechanical response and porous fibrillar structure.[10] The crystalline domains are attributed to the hydroxyl groups of glucopyranoses that form hydrogen bonds within and between cellulose chains leading to a helix formation. The poor mechanical performance and high hygroscopicity of cellulose may limit its applications in the packaging field. Modification of cellulose to obtain functional derivatives includes a two-stage chemical process. More specifically, alkaline treatment and sequential introduction of methyl- and/or hydroxypropyl-ether groups into the cellulose molecule, using chloroacetic acid, methyl chloride and/or propylene oxide are employed to produce carboxymethyl cellulose (CMC), methylcellulose (MC), hydroxypropylmethyl cellulose (HPMC) or hydroxypropyl cellulose (HPC).[10] These functional groups increase the hydrophobicity of cellulose derivatives while transparent cellulose ether-based films exhibit flexibility and moderate strength.[3,10] They show efficient O_2 and CO_2 barrier properties relatively comparable to synthetic packaging materials, such as dense polyethylene (HDPE) and low-density polyethylene (LDPE) films while they lag behind regarding water vapor barrier properties.[20] However, the intensity of modification significantly affects the biodegradability of these derivatives leading to decreased biodegradability.[10] Similarly to ether derivates, cellulose ester derivatives are obtained after acidic treatment followed by esterification with organic or inorganic acids to add ester groups to the cellulose molecule. The most common cellulose ester derivatives are cellophane, that is used in confectionery food packaging, and cellulose acetate.[10] Cellulose is a renewable, biocompatible, biodegradable and environment-friendly biopolymer and thus it has been extensively studied in various applications.[21,22] In 2018, the global cellulose market accounted for 219.53 billion USD and it is projected to grow with a combined annual growth rate of 4.2% by 2026 reaching 305.08 billion USD.[23] The recovery of plant cellulose requires pretreatment of the lignocellulosic feedstock to remove complex components bound to the cellulose structure, *i.e.*, hemicellulose and lignin. Conventional practice includes pulping (*via* acidic, alkaline or enzymatic hydrolysis), while bleaching using sodium chloride or hydrogen peroxide is also carried out through oxidating techniques.[2]

5.2.1.1.2 Chitin/Chitosan. Chitin is a polysaccharide that is abundantly found in animals as a structural element of their exoskeletons (*i.e.*, crustaceans, insects) as well as in cell walls of several fungal species. It is a linear homopolysaccharide composed of β-(1–4) *N*-acetyl-D-glucosamine units. The availability of chitin is rather restricted, with approximately 150 kilotons being intended for market use.[18] Alkaline or enzymatic deacetylation of chitin leads to the production of chitosan, which is composed mainly of β-(1–4) 2-amino-2-deoxy-D-glucose units. The incomplete deacetylation of chitin affects negatively the quality and properties of chitosan.[10]

The lower acetylation degree combined with higher nitrogen content of chitosan compared to chitin result in versatile properties, *i.e.* the development of nanocarrier systems and food packaging applications.[18,21] Chitosan is generally recognized as safe. Its antimicrobial properties are attributed to the cationic nature of the polymer and the degree of acetylation.[18,21] Neat chitosan-based films exhibit poor mechanical and barrier properties.[24]

5.2.1.1.3 Starch. Starch is a semicrystalline homopolysaccharide composed of linear (amylose) and highly branched (amylopectin) fragments. Amylose consists of α-D glucose linked *via* (1,4)-glycosidic bonds while in amylopectin α-(1,4)- (linear) and α-(1,6)- (branched) glycosidic bonds are observed. The ratio of amylose to amylopectin depends on the starch origin and this strongly affects the properties of starch-based films.[5] Starch can be found in abundancy in corn, wheat, potato and cassava with an amylose content ranging between 0 and 69%.[5,10] Due to its biodegradability and edibility, starch can be applied as an encapsulation agent in food and packaging sectors offering antioxidant and antimicrobial activity.[21] Acceptable tensile properties of starch-based films could be obtained *via* thermomechanical treatment of starch combined with plasticizers. However, fragility, poor moisture barrier properties and migration of plasticizers could limit its applications.[10]

5.2.1.1.4 Pectin. Pectin is a structural polysaccharide widely distributed in plant and fruit cell walls. It is biodegradable, biocompatible and non-toxic. The pectin domains found in the primary cell wall are homogalacturonan (HG), rhamnogalacturonan I (RG-I) and rhamnogalacturonan II (RG-II). They are mainly composed of linear chain D-galacturonic acid (GA) units.[25] HG consists of an α-(1,4)-linked-D-GA backbone partially methyl esterified, and constitutes 50–70% of the pectic material of the primary cell wall.[26] RG-I is a branched heteropolymer consisting of repeating α-(1,2)-L-rhamnose and α-(1,4)-D-GA units (linear backbone) with α-L-(1,5)-arabinan, β-D-(1,4)-galactan or arabinogalactan side chains. RG-II consists of a GA backbone linked *via* α-(1,4) bonds, while the four branched parts contain rhamnose, arabinose, galactose, glucuronic acid and other glycolyl residues.[26] The physical properties of pectin and pectin-based films are highly affected by the ratio of the HG, RG-I and RG-II domains.[25,26] Pectin can be classified as high-methoxyl pectin (HMP) (degree of esterification >50%) or low-methoxyl pectin (LMP) (degree of esterification <50%).[27] LMP forms a gel in the presence of calcium ions under pH values of 2–6, while HMP forms a gel when higher sugar contents and acidic conditions are applied. The molecular mass of pectin, diversity of branch chains or covalent interactions with other molecules, *i.e.* proteins, may also affect the film-forming capacity of the biopolymer.[25] Although pectin-based films present favorable gas barrier properties and delay lipid migration, they are brittle and lag behind conventional packaging concerning water vapor barriers. Neat pectin-based films have been reported to be fragile, exhibiting low tensile strength and high hydrophilicity.[28]

5.2.1.1.5 Marine Polysaccharides. Alginates are linear marine polysaccharides derived from seaweed species that have been utilized for the development of bio-based films due to their film-forming capacity.[13,29,30] Sodium alginates are composed of α-L-guluronic and β-D-mannuronic acid monomers linked to α-(1,4)-glycosidic bonds which offer good mechanical properties.[30] Alginate-based films present poor water barrier properties due to their hydrophilic nature.[29] Other algal polysaccharides that are used in bio-based film formulations include κ-carrageenan[31] and furcellaran.[32]

5.2.1.2 Proteins

Proteins are heteropolymeric and complex macromolecules consisting of polypeptide chains of various amino acids.[10] Polypeptide chains containing 50–2000 amino acids, with the latter being 22 in total (20 standard amino acids along with pyrrolysine and selenocysteine), are characterized as proteins.[21] Simple proteins are classified into globular (*i.e.* wheat gluten, corn zein) which are normally soluble in water, and scleroproteins which are fibrous and insoluble in water.[33] The physicochemical properties of proteins are affected by their structure and source of origin. Animal proteins are divided into hard and soft proteins.[34] Hard proteins, *i.e.* globular proteins, are colloidal molecules with high internal stability which present resistance to structural changes during adsorption. In the case of soft proteins (*e.g.* casein, gelatin) the structure is more disordered, with high susceptibility to internal changes.[34] Proteins are amphiphilic molecules and they tend to create absorbed interfacial layers between hydrophilic and hydrophobic regions.[35] Proteins have been widely used in food packaging preparations, including edible films and coatings[31,36,37] as well as active packaging materials.[38,39] Protein-based films exhibit superior mechanical and barrier properties compared to lipid- or polysaccharide-based films,[31,33] while their potential as substitutes for conventional petro-based packaging materials has deteriorated due to their having inferior water vapor barrier properties, high hydrophilicity and poorer mechanical properties. Bioactive proteins found in plant or animal tissues and microorganisms possessing antimicrobial and antioxidant properties have also been reported.[21] Antimicrobial proteins are gaining interest as alternatives to conventional antibiotics in pharmaceuticals and medicine. Recently, they have been applied as delivery systems in active food packaging applications as polymeric matrices that provide prolonged shelf life of products, simultaneously ensuring quality and safety.[40] Collagen, gelatin, whey proteins, corn zein, wheat gluten and soy protein are among the most commonly used in food packaging materials and edible films.[33]

5.2.1.2.1 Collagen/Gelatin. Collagen is the major structural protein of the extracellular matrix of connective tissues in animals.[12] It is a fibrous protein with a right-handed triple helix of left-handed parallel polypeptide chains linked *via* hydrogen and hydrophobic interactions. The cross-linked fibrils of collagen render collagen-based films and coatings insoluble in water.[12,41] Chemical treatment (acidic or alkaline) of collagen and sequential heating up to 40 °C leads to denaturation and gelatin production.[33] Gelatin is a complex water-soluble compound characterized by transparency and the absence of odor and smell.[33] The composition and gelling properties of gelatin are associated with the gelatin source (bovine-, porcine-, fish- or poultry-derived gelatin).[12] Gelatin has been evaluated in edible film formulations and coatings,[36] while neat gelatin-based packaging materials are incomparable with the synthetic ones mainly due to brittleness and water sensitivity.[10]

5.2.1.2.2 Milk Proteins. Milk proteins (casein and whey proteins) also have been evaluated for the development of alternative food packaging. Casein is soluble in water, and is the predominant (80%) protein in milk. Four subunits are found in its structure, namely αs1 (38%), αs2 (10%), β (36%) and κ-casein (13%).[33,42] Casein is available in the form of sodium or calcium caseinates derived as a by-product of the milk skimming process (20 °C, pH 4.6) after precipitation (pH 7) using NaOH or $Ca(OH)_2$ respectively.[33] Caseinates are favorable in film-forming applications due to their capability to form intermolecular hydrogen, and hydrophobic as well as electrostatic bonds.[42] Caseinates (Na^+ caseinates mainly) are less soluble in water compared to other proteins and thus they could be considered as alternatives to conventional packaging materials. Calcium caseinate-based films exhibit a more rigid structure compared to sodium caseinate-based films with superior barrier properties, due to calcium cations that promote cross-linking interactions. Sodium caseinate films present favorable optical and mechanical properties.[33] Strong interactions occurring between polar and nonpolar amino acids lead to shrinking phenomena upon the drying process of casein-based films, affecting their fragility.[42]

Whey protein fraction constitutes around 20% of milk proteins and contains several globulins, albumins and proteose peptones.[10,33] Impurities included in protein concentrates (*i.e.* lactose) affect the properties of the obtained films.[33] Whey is a by-product of the cheese manufacturing process obtained after casein precipitation of milk. Whey protein is rich in sulfur-containing amino acids and can be found in concentrate (50–80% protein) or isolate (90% protein) forms depending on its purity.[10,42] Films of high structural consistency could be obtained in cases in which denaturated (heating at 75–100 °C or alkaline treatment) whey proteins are employed.[42] Whey protein offers transparency, good gas/aroma and lipid barrier properties,[10] while poor water vapor barriers are observed due to their hydrophilic nature.[33]

5.2.1.2.3 Corn Zein. Corn zein is a prolamin derived as the by-product of corn processing and accounts for 45–50% of corn protein mass.[29,42] Corn zein is highly hydrophobic, soluble in polar protic solvents (methanol or ethanol) or dipolar aprotic solvents (acetone).[42] However, only water and ethanol or a mixture of these are acceptable for bio-film development in food packaging applications. Corn zein possesses antimicrobial properties and thermoplastic behavior and thus it can be processed in similar ways to conventional polymers used in food packaging.[29,33]

5.2.1.2.4 Soy Protein. Soy protein is a by-product derived from soybean oil manufacturing.[29] The composition of soy protein includes fractions of albumins and globulins.[43] Soy proteins are commercially available in different forms including soy protein isolate (SPI; protein content ≥90%), soy protein concentrate (SPC; ~70%) and soy flour (~56%), while their carbohydrate content ranges between 2–34%.[29,33] Soy proteins are hydrophilic and sensitive to environmental changes resulting in weak barrier properties of the produced films. Soy protein-based films are superior to polysaccharide- or lipid-based films in terms of gas barrier properties, with the latter being quite comparable to low-density polyethylene films, starch- and pectin-based films when limiting moisture conditions are employed.[33] Soy protein is a cheap renewable resource in packaging film applications or coating matrices, with SPI being the major form applied due to its high purity. These applications are restricted due to competition with the food chain. Increased demand for soy protein for food consumption has been attributed to its high nutritional value.[43]

5.2.1.2.5 Wheat Gluten. Wheat gluten is a storage protein composed of high-molecular-weight glutenins and single-chain gliadins of lower molecular mass. Glutenin subunits are interlinked *via* disulfide bonds, while the single chain of gliadins is arranged *via* intramolecular disulfide bonds.[44] The viscoelastic profile of wheat proteins depends on the ratio of glutenins to gliadins. Neat gluten-based films produced without plasticizing agents lack flexibility because gluten is rich in glutamic acid and thus hydrogen bonding interactions are promoted between the protein chains. Gluten presents good film-forming properties and could be processed similarly to conventional plastics.[43]

5.2.1.2.6 Sunflower Protein. Sunflower seed constitutes a favorable source of protein. The global production of sunflower seed is estimated to be 21 796 million tonnes for the period 2021/2022.[45] Sunflower proteins can be isolated through alkaline extraction from sunflower meal which is obtained as the by-product of the oil manufacture process. It exhibits a protein content of approximately 30–40% depending on the variety and culture conditions.[46] Sunflower protein is obtained in the form of concentrates (SFPC) and isolates (SFPI) and it mainly includes globular proteins, *i.e.* helianthinins and albumins at a ratio of 2 : 1.[44] Based on the Osborne classification, globulins are insoluble to water though they are soluble in aqueous salt solutions, while albumins are water soluble.[34] Helianthinin is the main protein in sunflower, and has a high molecular mass (300 kDa). It consists of six subunits that contain acidic and basic side chains linked by disulfide bridges.[34,44] Albumins are low-molecular-weight molecules (up to 18 kDa) which consist of amino acids with basic side chains.[44] Based on the protein extraction process, the produced protein may contain various antioxidant compounds naturally occurring in sunflower seeds. Phenolic compounds, which are responsible for the intense color of sunflower proteins, are difficult to entirely remove from proteins due to the strong interactions between these two molecules.[47] The barrier properties and mechanical behavior of sunflower protein-based films have been reported to be similar to other protein-based films of different origin (*i.e.* SPI).[48] However, the color intensity observed in sunflower protein-based films is highly associated with consumer acceptability issues and thus recent scientific studies dealing with sunflower protein are limited. Opacity issues of sunflower protein-based films related to whole package applications have shifted interest toward specific practices, *i.e.* sealing packaging membranes, that could additionally serve as carrier matrices for several antioxidants, vitamins and antimicrobial agents. This ability combined with phenolic compounds naturally occurring in sunflower protein could sustain active packaging applications adding value, quality and extended shelf life to fresh or organic edible products. Sunflower protein-based membranes enriched with antimicrobial agents have been reported to delay lipid auto-oxidation and growth of the total mesophiles during fish product preservation.[47]

5.2.1.3 Lipid-based Biopolymers

Lipid-based polymers include molecules of non-polar and hydrophobic nature like fats, oils and waxes. Fats and oils are esters of fatty acids and glycerol presenting solid-like or liquid-like behavior depending on their saturation and unsaturation fracture, and they are derived from plants, animals or microbial sources. Waxes composed of esterified long-chain alcohols with fatty acids may be natural or synthetic, depending on their origin.[29] Lipid-based biopolymers are mainly used as reinforcing agents in composite materials to improve the properties of films and coatings.[10] The hydrophobic nature of lipid-based polymers may lead to enhanced barrier and thermal stability of the final formulations[49–51] as well as enhanced quality (unsaturated and essential fatty acids) and extended shelf life of packed edible products.[13,50,51]

5.2.2 Synthetic Polymers Derived from Bio-based Monomers

PLA is a bio-based aliphatic polyester composed of lactic acid building-blocks derived through microbial bioconversion of carbohydrates from renewable resources (*i.e.*, sugarcane or sugar beet).[10,21] PLA is prepared through polymerization processes of D- or L-lactic acid enantiomers including polycondensation, azeotropic polycondensation, solid-state polymerization and ring opening polymerization,[52,53] with direct polycondensation and ring opening polymerization being the most widely used. PLA polycondensation includes the bonding of lactic acid monomers *via* reactions occurring between carboxyl and hydroxyl groups with simultaneous removal of water by-products.[54] The ratio of D- and L-lactic acid enantiomers as well as the polymerization method employed are highly associated with the physicochemical properties of the final PLA formulation. Direct polycondensation of lactic acid leads to the formation of low-molecular-weight ($<50\,000$ g mol^{-1}) PLA chains with high fragility, while they require further modification to be valorized as packaging material.[53] Ring-opening polymerization is a chain growth process of lactides with ionic reactive centers. Lactides are cyclic dimers of lactic acid resulting from low-molecular-weight PLA-polycondensates through depolymerization.[10] Through ring-opening polymerization PLA of high molecular weight is generated without further modifications and thus this process is the most commonly used.[53] The crystalline structure of PLA is highly associated with the percentage of L- and D-enantiomers. PLLA and PDLA are semi-crystalline, while PDLLA (mixture of PDLA and PLLA) is amorphous.[10,52] PLA is a transparent, thermoplastic bio-based polymer with enhanced mechanical properties that can be processed similarly to conventional fossil-based materials and thus it has been used in packaging film and container applications.[10,52] PLA-based films exhibit excellent barrier properties compared to conventional packaging materials, however, they are brittle with low elongation at break.[10,55] To overcome these drawbacks the addition of plasticizing or reinforcing agents has been proposed.[10,55,56]

5.2.3 Microbial Biopolymers

Microorganisms are able to produce intra- or extracellularly numerous biopolymers with diverse biological functions (*i.e.*, energy storage, protection) as a response to environmental changes. Among these, two major fermentation-based polymers are polyhydroxyalkanoates (PHA) and exopolysaccharides. Crude renewable resources could be valorized as carbon and nutrient sources through microbial bioconversion to obtain PHAs or exopolysaccharides.[57]

5.2.3.1 Poly(hydroxyalkanoates) (PHAs)

PHAs are linear polyesters of hydroxyalkanoates produced as secondary metabolites by numerous bacteria, among which *Cupriavidus necator* and *Ralstonia eutropha* are the most commonly used.[58] PHAs are synthesized *via* acetyl CoA conversion to *R*-3-hydroxyacyl CoA and further polymerization,[58] under growth-limiting fermentation conditions involving limitation of oxygen, nitrogen or phosphorus while control of the fermentation parameters can lead to high accumulation of these polymers (up to 90% of the microorganism dry mass).[10] These polyesters are assembled intracellularly into hydrophobic granules and serve as carbon and energy sources for microorganisms.[8] Based on the monomer (3-hydroxyalkanoates) type, PHAs are distinguished into short-chain-length (SCL) PHAs consisting of monomers with up to five carbon atoms and medium-chain-length (MCL) PHAs including 3-hydroxyalkanoates consisting of six to fourteen carbon atoms.[58] PHAs are optically active thermoplastic elastomers exhibiting sufficient UV resistance and high molecular weight. However, their inferior mechanical behavior compared to conventional packaging materials, as well as their susceptibility to thermal degradation (~180 °C) and slow solidification, limit PHA's applications and make their processing challenging.[10,59] The properties of PHAs are highly dependent on the physicochemical properties of 3-hydroxyalkanoates, fermentation conditions and bacterial strains applied,[58] while the production of PHAs with tailored properties meeting specifications of industrial applications may be obtained through bioengineering of bacterial species, as well as physical blending and chemical modification.[8] The main representative of PHAs is poly(3-hydroxybutyrate) (PHB). PHB is classified as a SCL-PHA and is highly crystalline. Favorable water vapor and aroma barrier properties make it an ideal candidate for short-term food packaging applications.[58] However, high brittleness along with low thermal stability of pure PHB lead to deteriorated mechanical profile.[59] To develop PHB-based materials with improved properties several approaches have been proposed including blend formulations and copolymerization.[60–62] PHAs rank among the most expensive bio-based plastics, mainly due to energy demands for their extraction and purification.[10] Current advances in bioprocessing within the concept of circular economy and biorefinery scenarios could enable PHB cost-competitiveness.[63]

5.2.3.2 Bacterial Cellulose

Bacterial cellulose (BC) is an extracellular polymer produced by numerous bacteria including species of *Komagataeibacter* (formerly *Gluconoacetobacter*), *Rhizobium*, *Agrobacterium* and *Sarcina*, while bacteria of the *Komagataeibacter* genus are among the most efficient BC producers.[1,58]

BC has been extensively studied and characterized due to its superior inherent properties compared to plant-based cellulose, including absolute purity, high degree of polymerization and crystallinity as well as biocompatibility.[2,64] BC synthesis involves polymerization of glucose monomers to β-1,4-glucan linear chains excreted in the cytoplasm as nanofibrils. Consequently, nanofibers assemble into microfibrils *via* crystallization leading to the formation of microfibrillar bundles to form a 3D BC network.[58] The average diameter and length of the BC fibrils in the BC structure are 100 nm and 100 μm, respectively.[2] BC is obtained as a gel-like membrane, while the high aspect ratio of the fibrils results in a high surface area which induces high absorption of water and strong interactions with other biopolymers and nanocompounds.[65]

The unique properties of BC, *i.e.* enhanced mechanical behavior (tensile strength and Young's modulus) and high thermal stability offer a broad spectrum for novel applications from nanomaterial synthesis to the food, packaging and textile industries as well as the biomedical and pharmaceutical fields.[58] The properties of BC are highly dependent on factors such as fermentation conditions, cultivated strain, raw materials used as substrate as well as downstream processing and post-harvesting operations.[2] NaOH-treated BC membranes have been reported to present two-fold higher Young's modulus compared to untreated membranes.[1] *In situ* modification of BC *via* the addition of functional molecules, *i.e.* polysaccharides, proteins and cellulose derivatives, to the fermentation medium is an effective way to achieve desirable BC properties. Substitution of glucose with bamboo enzymatic hydrolysate as a carbon source combined with the addition of sodium alginate during BC biosynthesis from *A. xylinum* ATCC23767 has been reported to enhance the thermal properties of the obtained BC-based composites compared to native BC.[66]

The multiple hydroxyl groups present in the BC structure allow the implementation of modifications leading to value-added composites with tailor-made properties for targeted applications.[67] Cross-linking has been effectively employed to prepare BC-based composites with improved mechanical properties using cross-linking agents such as glutaraldehyde. Yang *et al.*[68] demonstrated that BC-based composite films obtained after cross-linking with glutaraldehyde or 1-ethyl-3-(3-dimethylaminopropyl) carbodiimide hydrochloride exhibited enhanced thermal stability and improved tensile strength by 57.9% and 70.8%, respectively, compared to BC.

Wang *et al.*[69] employed *ex situ* modification of BC produced by *A. xylinum* ATCC 1765. Their results indicated that *ex situ* modified BC fibers presented enhanced tensile strength and Young's modulus compared to BC fibers that were alkaline and wet-twisting treated. Moreover, *ex situ* modification involving hydrolysis (acid or enzymatic) of BC membranes has successfully being employed to obtain high-performance nanomaterials that could be used as reinforcing agents in film polymeric matrices.[14,70]

BC presents excellent mechanical behavior, gas and moisture barrier properties, as well as thermal stability compared to other polysaccharides, rendering it an appealing alternative to conventional packaging materials. However, the limited elasticity of BC (1.10–6.57%) is the major drawback hindering its handling.[71]

5.3 Natural and Microbial Biopolymers in Food Packaging

Food packaging is a large growing sector of the plastic packaging field. In 2020, food and beverage packaging led the worldwide plastic market with a

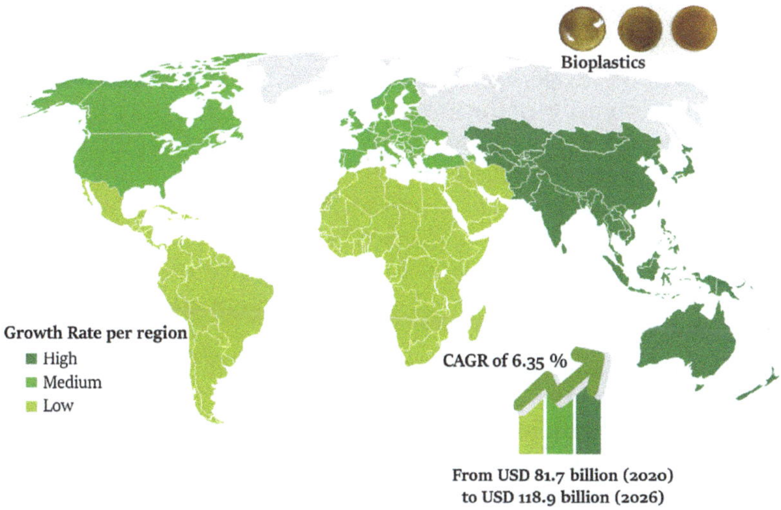

Figure 5.2 Market potential of bioplastics by region, forecast period 2021–2026.[180]

revenue share of 51%,[72] while the food packaging market is forecast to grow with a compound annual growth rate (CAGR) of 6.2% for the period 2022–2030, reaching 466.9 billion USD by 2027.[73] Rising environmental and societal concerns interlinked with waste generation, decrease of primary resources, sustainability and food safety issues as well as lifestyle changing of contemporary consumers have gradually increased interest toward biodegradable polymers that can be found in abundance, and offer biocompatibility and renewability. Bio-based packaging materials are gradually replacing conventional fossil-based constituents with biodegradable packaging market projected to reach 118.85 billion USD by 2026, with a CAGR of 6.3% between 2021–2026[74] (Figure 5.2). Food quality preservation and food safety assurance are priority issues of packaging materials. Novel bio-based packaging (edible films/coatings) could be applied in various food products preventing quality deterioration and providing physical protection combined with water-loss reduction, restriction of O_2 absorption and limiting lipid oxidation.[35] Bio-based composites and biopolymeric blends have presented remarkable functional properties including water and gas barrier behavior. The incorporation of natural colorants and pigments in the final formulations has been applied to develop novel and intelligent packaging allowing the consumer to evaluate the quality of the product on site.[29]

5.3.1 Preparation Techniques and Strategies of Biofilms

The production of biopolymer-based packaging materials is a multiphase process involving disruption of intramolecular bonds, rearrangement of the molecules and synthesis of a 3D structural network stabilized by noncovalent and covalent interactions.[57]

Two main approaches have been proposed for the development of bio-based films, namely wet and dry processing.[10] Wet processing or solvent casting involves the dispersion of film-forming biopolymers in solvent systems, pouring or spraying of the mixtures in a mold surface and finally drying under mild temperature conditions to produce films.[57,75] Coatings of edibles are also produced *via* direct application of the mixture on the product surface while the type of product surface determines the method employed. Dipping is preferred in the case of irregular-shaped products, while spraying is applied over large surface areas.[10] The type of solvent and pH during wet processing are important factors that affect the polymer conformation.[57] This method normally employs green solvents, *i.e.* water (pure or acidified) and solvent systems including ethanol.[12] Wet processing is widely applied in lab-scale production, however, the drying step involving mild temperatures, limits industrial-scale applications due to time-consuming issues.[75]

Dry processing is associated with the thermoplastic behavior of biopolymers and has been related to the semi-solid conversion of glassy materials under specific temperatures. This semi-solid condition disorders the polymer structure causing molecular mobility, breakage of intramolecular bonds and the formation of new linkages leading to modified physicochemical and mechanical properties of the final material.[57] Dry film processing is a short cycle method that could be easily transferred to large-scale manufacturing. It is a solvent-free method with low environmental impact and includes hot pressing, extrusion and compression molding.[12] Compression molding (cold or hot press) is a fast, low-cost method with high-volume production capacity.[75] However, extrusion is the most common process currently employed in the manufacture of synthetic plastics. Extrusion is a flexible process permitting handling of a broad range of materials. Physicochemical alterations of the material due to processing conditions as well as high-pressure drop in the extruder die lead to modified properties of the final formulations.[75] Jiménez-Rosado *et al.*[76] prepared wheat gluten-based composites *via* extrusion and hot compression molding methods. The authors reported that extrusion led to bioplastic sheets of higher compatibility between their components, compared to compression molding. This was probably attributed to the greater level of aggregation-cross-linking and structural alignment within the produced sheets in the flow direction of the extrusion. The extruded bio-based sheets presented improved mechanical and water-uptake properties.

Bio-based films derived from a sole biopolymer as matrix usually present a brittle structure and inferior mechanical properties compared to conventional packaging materials.[75] Thus, the incorporation of plasticizers could increase the functionality and flexibility of the final formulations. Plasticizers reduce polymer chain stiffness, altering the polymer–polymer interactions in the bio-polymeric matrix. This leads to an increase in the free volume and chain mobility, resulting in high fracture resistance of the final bio-based materials.[12,57] Various substances have been utilized as plasticizers including glycerol, ethylene glycol, oils and waxes and others.[12]

Sufficient plasticization relies on the ratio of biopolymer to plasticizer as well as the molecular weight and polarity of the plasticizer. Polar plasticizing agents decrease hydrogen interactions, resulting in greater disruption to the biopolymer matrix and thus more effective plasticization.[77] Thus, selection of the appropriate plasticizers depending on the nature of the bio-based matrix is crucial to preventing insufficient plasticization. Indicatively, amphiphilic plasticizers favor plasticization of corn zein protein-based films, while hydrophilic substances, *i.e.* glycerol, would only affect the hydrophilic regions leaving intact the hydrophobic core of the protein.[78]

5.3.2 Physicochemical Properties

The physicochemical properties of biopolymer films are crucial to their function. Biopolymer-based formulations, including a neat biopolymer as the sole film-forming component, show poor mechanical behavior, water resistance, and water and gas barrier properties. Polysaccharide-based films present sufficient mechanical strength and elasticity, while they lag behind in terms of water barrier properties compared to synthetic polymers.[79] Protein-based films usually present excellent gas barrier properties (500 percent lower oxygen permeability compared to polyethylene has been reported in some cases) and sufficient mechanical properties compared to other biopolymer-based films. However, poor water resistance and water vapor permeability are major drawbacks for their widespread applications.[42]

Table 5.1 summarizes variations in physicochemical properties of various bio-based films produced *via* different techniques based on recent publications. Investigation and optimization of physicochemical properties of bio-based films are challenging and various strategies have been proposed to improve the functional properties of bio-based films including plasticization,[80–82] blends of biopolymeric substances,[24,31,36,83,84] incorporation of nanofillers[85–91] or active ingredients.[32,37,38,92,93] In some cases further modifications of the formulations, i.e. employing cross-linking[81] with transglutaminase,[94] are required to obtain functional properties comparable to conventional packaging materials.

The mechanical properties of food packaging films provide straightforward information on their performance in handling and storage. Hydrogel films were prepared by ternary blends of κ-carrageenan, xanthan and gellan gum by solution casting based mainly on hydrogen bond interactions between the polymeric chains.[95] Their elongation at break was 14–19%, the tensile strength was 20–30 MPa and their tensile modulus about 2 GPa. These values decreased by reducing the content of any of the polysaccharides. The enhanced mechanical properties of the films were attributed to their semicrystalline nature. In the study of Salari *et al.*[87] neat chitosan-based films exhibited tensile strength and Young modulus of 21.1 MPa and 97.0 MPa, respectively, with water vapor permeability (WVP) of 3.65×10^{-10} $g\,m^{-2}\,s^{-1}\,Pa^{-1}$ and 22.4% solubility in water. Enhanced properties of the bio-based films by 43.7% and 15.6% for Young modulus and

Table 5.1 Physicochemical properties of various bio-based films that are produced employing various bio-polymers based on recent scientific studies.[a]

Composition	Additional agents	TS (MPa)	EB (%)	YM (MPa)	WS (%)	MC (%)	SD (%)	WVP ($g\,m^{-2}\,s^{-1}\,Pa^{-1}$)	ΔE^*	T_d (°C)	T_g (°C)	T_m (°C)	Ref.
Casein	—	2.6	458	—	100	24.4	—	11.00×10^{10}	—	308	55.4	—	81
	20% tannic acid	~5.8	~290	—	23.7	11.9	605.6	4.80×10^{10}	—	323	67.4	—	
Chitosan	—	14.2	64.9	—	—	—	—	1.86×10^{16}	91.3	230	—	—	24
Chitosan/zein	—	19.1	72.8	—	—	—	—	1.22×10^{16}	91.4	240	—	—	
Gelatin	—	41.0	13.3	1596	—	—	—	3.19×10^{-11}	—	—	22.3	82.0	37
	0.38% MEO[b]	29.0	11.1	1047	—	—	—	4.06×10^{-11}	—	—	22.0	83.4	
	0.50% MEO[b]	27.0	10.0	1048	—	—	—	4.19×10^{-11}	—	—	27.0	83.1	
Zein	—	11.3	5.67	346.6	—	—	—	1.37×10^{-10}	29.1	—	—	—	38
Gelatin	—	14.3	104.4	109.4	—	—	—	2.82×10^{-10}	3.9	—	—	—	
Multilayer zein/gelatin	Tea polyphenols	4.6–11.1	9.9–49.8	120.6–276.7	—	—	—	1.70×10^{-10}–3.00×10^{-10}	22.0–9.5	—	—	—	
Tapioca starch	10% CNF	18.3	5.6	—	16.4	11.8	—	—	2.4	308	—	151.2	88
Potato starch	10% CNF	17.5	8.3	—	16.8	11.0	—	—	2.4	298	—	165.0	
Gelatin	—	6.2	11.0	—	—	—	—	2.05×10^{-10}	—	—	69.6	—	36
Gelatin/pectin	—	14.4	4.4	—	—	—	—	1.54×10^{-10}	—	—	78.9	—	
Pectin/gelatin	—	9.5	41.8	205.4	—	—	—	4.44×10^{-11}	—	—	—	—	91
Pectin/gelatin	HT[c]/DHPG[d]	16.6	53.4	287.6	—	—	—	3.61×10^{-11}	—	—	—	—	
Furcellaran/WP[e]	—	—	26.0	198.0	47.4	14.3	285.6	—	—	—	—	218.2	32
	Yerba mate extract	—	22.2	238.1	46.9	12.0	336.1	—	5.1	—	—	219.5	
	White tea extract	—	25.1	214.1	51.9	14.1	265.9	—	1.4	—	—	219.4	
Chitosan/quinoa protein	—	6.9	36.9	—	—	—	—	4.75×10^{-9}	—	—	—	—	92
	1% thymol	7.8	40.2	—	—	—	—	3.94×10^{-9}	—	—	—	—	
Chitosan/protein	—	11.6	55.7	—	—	—	—	2.78×10^{-10}	—	—	—	—	85
Chitosan/protein/linseed oil	—	8.5	40.6	—	—	—	—	3.25×10^{-10}	—	—	—	—	
	ZnO	17.7	32.7	—	—	—	—	2.83×10^{-10}	—	—	—	—	
SP[f]	—	1.9	55.8	5.5	37.4	33.5	~1500	1.56×10^{-10}	—	—	—	—	82
SP/galactomannan	—	3.7	38.0	7.1	58.7	30.7	~800	1.20×10^{-10}	—	—	—	—	
Zein protein	—	2.3	1.5	—	14.7	22.0	104.8	1.40×10^{-10}	—	—	—	—	90
	100% CINPs[g]	12.5	4.5	—	14.6	20.1	17.2	8.08×10^{-11}	—	—	—	—	
Pectin	—	7.1	~20	—	—	—	—	1.28×10^{-10}	—	—	56.6	167.5	89
	5% CNC	13.1	~30	—	—	—	—	9.30×10^{-11}	—	—	57.5	169.1	

Pectin	1% $CaCl_2$	14.8	6.4	2.3	25.4	18.4	3.4	13.22×10^{-11}	6.8	228.9	—	228.9	93
	1% $CaCl_2$/1.5% CEO[h]	33.8	11.7	2.9	17.4	10.1	1.9	6.52×10^{-11}	10.5	232.2	—	232.2	
WP[e]	—	~5.5	~19.8	~1.6	70.6	19.6	—	2.13×10^{10}	1.6	—	—	—	83
	3% chitosan	~4.8	~12.5	~1.7	51.6	18.0	—	2.29×10^{10}	6.7	—	—	—	
EWP[i]/κ-C[j] (80 : 20)	—	7.4	8.6	—	—	—	—	1.59×10^{-3}	82.3	—	—	—	31
EWP[i]/κ-C[j] (40 : 60)	—	30.0	10.8	—	—	—	—	2.12×10^{-3}	84.0	122.6	—	—	
κ-C[j]	—	36.7	9.3	—	—	—	—	2.55×10^{-3}	73.8	164.9	—	—	
PLA	—	41.1	139.6	—	—	0.08	—	1.69×10^{-11}	—	338.6	35.5	134.9	84
	4% CMC	162.5	177.9	—	—	0.61	—	2.60×10^{-11}	—	303.0	25.9	126.1	
SP[f]	—	5.7	38.4	—	31.3	—	—	1.68×10^{10}	17.7	—	—	—	94
SP[f]/SC[k]	—	10.8	17.8	—	48.4	—	—	1.43×10^{10}	9.0	—	—	—	
	Transglutaminase	16.4	10.2	—	19.8	—	—	1.14×10^{10}	9.1	—	—	—	
Chitosan	—	21.1	33.8	97.0	22.4	18.5	—	3.65×10^{10}	17.5	—	61.7	111.1	87
	1% AgNPs	24.3	33.0	139.5	21.6	18.0	—	3.31×10^{10}	19.4	—	62.3	116.7	
	4% BCNC	41.3	23.8	297.4	19.4	14.6	—	2.68×10^{10}	15.8	—	74.3	123.4	
	4% BCNC/1% AgNPs	42.9	22.5	311.9	16.9	11.7	—	2.16×10^{10}	18.6	—	75.3	125.1	

[a]Tensile strength (TS); elongation at break (EB); Young's modulus (YM); solubility in water (WS); moisture content (MC); swelling degree (SD); water vapor permeability (WVP); color difference (ΔE^*); decomposition temperature (T_d); glass transition temperature (T_g), melting temperature (T_m).

[b]MEO: mint essential oil.

[c]HT: hydroxytyrosol.

[d]DHPG: 3,4-dihydroxy-phenylglycol.

[e]WP: whey protein.

[f]SP: soy protein.

[g]CINPs: catechin/β-cyclodextrin inclusion complex nanoparticles.

[h]CEO: clove bud essential oil.

[i]EWP: egg white protein.

[j]κ-C: κ-carrageenan.

[k]SC: sodium caseinate.

tensile strength, respectively, were achieved *via* incorporation of 1% silver nanoparticles (AgNPs) while solubility in water and WVP were decreased by 3.5% and 9.2%, respectively. The authors reported further improvement of the film's functional properties *via* the incorporation of bacterial cellulose nanocrystals (BCNs) in the final formulations.

Rheological measurements are often applied on biopolymer solutions that are going to be used for film preparations. The concentrations of pectin and sodium alginate in pectin/sodium alginate/xanthan solutions were optimized by steady flow experiments.[30] The power-law behavior of the viscosity as a function of shear rate was investigated and found to be related to the preparation of the final films. A balance was sought where the solutions could form free-standing films while these films were still homogeneous. In addition, linear rheology measurements were performed to spot the concentration were a crossover in the storage and loss modulus was observed at high frequencies. In this situation the overlap and cross-linking of chains supported the formation of films with good mechanical properties.

Strong hydrogen bonding biopolymers give rise to films with high susceptibility to water, while polymers with high hydrophobicity lead to final formulations with great moisture barrier properties. Water vapor permeability is greatly affected by polymer–water interactions and the ratio of hydrophobic/hydrophilic counterparts.[35] Water sorption and permeation highly affect the physical and chemical properties of the food packaging systems. Thermodynamic analysis of water sorption in whey protein isolate/polyvinyl alcohol blends showed that a water activity lower than 0.75 was necessary for stable films.[96] Water vapor permeability is related to the barrier properties of the films and has been observed to decrease with cross-linker addition.[97] The swelling ratio in water is a measure of the ability to absorb water and is related to the mechanical properties in the highly hydrated state. In starch hydrogel films the addition of citric acid reduced swelling because of ester bonds and cross-linking, leading to higher integrity.[98]

Thermal gravimetric analysis is customarily used to characterize the stability of the biopolymer films. There are several zones of weight loss as a function of temperature.[98] For example, in the antioxidant starch films cross-linked with citric acid, evaporation of water occurs up to 150 °C and starch degradation takes place between 150 and 270 °C. Starch decomposes up to 350 °C, while at higher temperatures the residual weight is measured. Differential scanning calorimetry provides important insights into the thermal transitions in packaging materials. The glass transition temperature, melting temperature and melting enthalpies are measured. In collagen/laponite biocomposite films the behavior of partially crystalline materials was observed.[99] In more detail, a first glass transition was measured between −76 °C and −72 °C and was caused by the fraction which was rich in glycerol. The second glass transition, between 42 °C and 43 °C was attributed to the collagen-rich phase. It was pointed out that in the temperature range between 0 °C and room temperature (below glass transition) the films will have better mechanical strength and water vapor barrier properties.

X-ray diffraction is extensively used to characterize the crystalline nature of films based on biomacromolecules. In κ-carrageenan/nanocellulose nanocomposite films strong intensity peaks at angles 2θ between 15° and 34° revealed the crystallographic planes of cellulose.[100] The relative intensity of a characteristic crystallographic peak in relation to the intensity between certain peaks provided a crystallinity index in the order of 70–90% which was attributed to the chemical treatment during the production of the cellulose nanocrystals in which hemicellulose and lignin were removed. Films with high crystalline content are expected to have increased mechanical properties, *i.e.* stiffness and tensile strength. Very recently, edible films based on carboxymethyl chitosan and pullulan that were enriched with galangal essential oil were proposed as eco-friendly packaging material.[101] The arrangement of pullulan chains was tight and represented by broad amorphous peaks. On the other hand, X-ray diffraction patterns of carboxymethyl chitosan were sharp owing to its crystalline organization. In carboxymethyl chitosan/essential oil mixtures a broad amorphous peak is found. In the blend films that contain all three components, an amorphous diffraction peak was observed, which was shifting to lower angles as more pullulan was added. This was attributed to the interaction between the two polysaccharides.

The mesoscopic morphology of biopolymer films can be investigated by small angle scattering methods (SAS).[102,103] These techniques are suitable for resolving the spatial characteristics at length scales between 1 and 100 nm or higher. Small angle neutron scattering (SANS) has the advantageous capability of contrast variation and contrast matching, where specific parts of a sample can be highlighted by the use of the proper combination of normal and deuterated molecules. Small angle X-ray scattering (SAXS) has the advantage of fast measurement and very high angular and wavelength resolution. Both methods are non-invasive and provide averages over macroscopic sample volumes.

A very general description of a small angle scattering profile from a hierarchically organized nanocomposite film is illustrated in Figure 5.3. At low q (large length scales) the characteristic power-law $I(q) \sim q^{-m}$ corresponds to the interface or fractal internal cluster arrangement of large objects whose size falls outside the q-measurement window. At higher q the scattering from macromolecular clusters of gyration radius $R_{g,clust}$ may be observed. At $q > 1/R_{g,clust}$ the internal structure of the clusters is measured. The characteristic exponent n may be the one of a mass fractal ($1 < n < 3$). At $q > 1/\xi$ the correlations of the individual macromolecular structures are observed by the fractal exponent α. For example, in the case of rigid cylinders or rods, $\alpha = 1$. The characteristic size ξ is the mesh size, *i.e.* the mean distance between different structures. At q higher than the transverse size of the structures R_0 the form factor of the cross-section dominates the scattering. At even higher

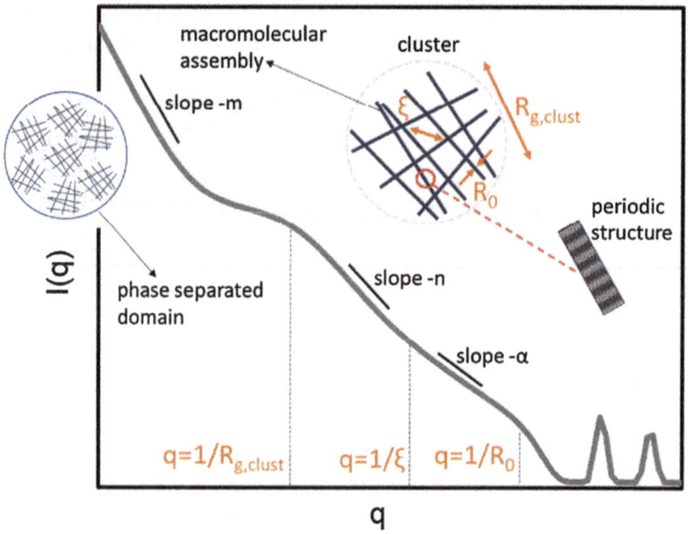

Figure 5.3 Illustration of an SAS profile from a hierarchically organized bionanocomposite film.

length scales a possible periodic structure would be observed as Bragg peaks in the scattering function.

Konjac glucomannan/agar binary films of thickness between 60 and 95 μm were prepared by casting mixed solutions of the two components.[104] The authors attributed the characteristic features of the SAXS patterns at high scattering wave vectors (small length scales) to the presence of crystallites within the amorphous matrix. The increase of the konjac glucomannan content resulted in an increase in the related scattered intensity due to better packing within the ordered regions. However, at very high contents the amorphous phase dominated and the scattered intensity decreased. Modified proteins from fish with gelling ability were used for biopackaging films.[105] The SAXS data were modeled by a function that contained a power-law term at low q and a term at intermediate and high q that defined the correlation length of the gel. The morphology at large length scales (low q) was the one of mass fractal as it was defined by the characteristic power-law exponent that was about 2. This fractal structure was attributed to the protein aggregation after denaturation. The correlation length was higher at pH 10 in comparison to pH 2 as the mechanism of denaturation was accompanied by disulfide bond splitting.

Pure lignocellulosic fractions were obtained by *Posidonia oceanica* residues and were used in composite starch films.[106] In the SAXS profiles, a broad shoulder was observed at intermediate scattering wave vectors (q) which in real space corresponded to about 10 nm. This was attributed to the heterogeneities of the semicrystalline structure of starch. At low q a power-law behavior $I(q) \sim q^{-m}$ was found. This was related to the disrupted lamellar structure (due to extrusion) in the case of pure starch ($m = 1.5$). The addition of lignocellulosic fractions led to exponents *m* near 4, which corresponds to sharp interfaces and indicates two-phase systems. These findings supported the results of this work from WAXS and water permeability. Cellulose nanocrystals were incorporated in chitosan matrices to prepare nanocomposite films.[107] Neutralization of the films was performed and the effects on the water resistance and antibacterial properties were investigated. The width of the cellulose nanocrystal aggregates was estimated by SAXS. The $I(q)q^2$ curves showed a maximum whose position q_{max} was translated into width by $L = 2\pi/q_{max}$. It was found that this size increased as the concentration of cellulose nanocrystals increased. Thermoplastic corn starch films with microalgae species were proposed as biodegradable materials for food packaging.[108] A broad shoulder at *q* corresponding to 9–10 nm was connected to the lamellae in the starch granules which are heterogeneously distributed in melt blended starches. In the presence of microalgae the water permeation was restricted and the starch chains were not able to pack into lamellae. This way, the shoulder in the SAXS data was lost.

5.3.3 Nanostructure

Alternative biopolymer-based packaging is composed of non-toxic, biodegradable and biocompatible components.[21] The enhancement of barrier and mechanical behaviors of bio-based packaging materials is of great interest as drawbacks interlinked with thermal stability, brittle structure and inadequate gas and water barrier properties of the final formulations limit their wide applications. Within this context, nanotechnology could be a useful tool to develop novel alternative packaging with sufficient properties and performance. Incorporation of active nanostructures (antioxidant and/or antibacterial) into biopolymeric matrices could improve their functionality and bioactivity.[90]

Nanocomposite materials include a wide spectrum of substances composed of more than two compounds, of which at least one presents nanoscale dimensions (1–100 nm). The incorporation of homogeneous nanoparticles, characterized by a high aspect ratio, increases the matrix–filler interfacial area, leading to diverse thermal and mechanical responses.[89] Also, the newly created biopolymer–nanoparticle linkages lead to a reduced number of biopolymer chain sites that are available to interact with water molecules affecting the water barrier behavior of nanocomposites.[109] Various nanoparticles have been utilized to develop novel nanocomposite materials with desirable functional properties including organic and inorganic components.

The use of zinc oxide (ZnO) as the component phase of nanocomposite materials to improve targeted functional properties as well as the application of final nanocomposite materials on edible products including fresh fruit and vegetables, meat and bread products has been extensively proposed.[19,85,110] The antimicrobial functions of ZnO nanocomposite materials may be attributed to the potential emersion of reactive oxygen bondings

(-OH, O_2^-, H_2O_2) with biological macromolecules, the antibacterial properties of Zn cations, and finally to the microbial cell damage caused by ZnO *via* increase of the microbial membrane permeability.[109]

Noshirvani *et al.*[19] suggested chitosan–CMC–oleic acid-based films and coatings incorporated with ZnO nanoparticles (ZnONps) as alternative packaging materials for wheat bread products. The authors concluded that incorporation of ZnO NPs into the polymeric-based matrix resulted in significantly reduced water vapor permeability of the prepared nanocomposites compared to control films. Extended shelf life of bread products was observed during 35 days of storage compared to control bread slices without film or coating packaging. More recently, Wang *et al.*[85] prepared chitosan–protein–linseed oil films reinforced with ZnO NPs and evaluated their performance on the quality of raw meat products. According to their results, the addition of ZnO NPs led to sufficient improvement of the mechanical properties as well as water vapor barrier behavior of the final formulations, while application of the films as wrapping packaging of raw meat products indicated great sensory properties and reduced the total bacterial counts.

Other oxides such as silica,[111] nanoclays, *i.e.* montmorillonite[112] or organic nanoparticles including nanocellulose[113] and nanostructured chitosan particles,[92] have also been employed as reinforcing agents of packaging materials improving the functionality of the final formulations. An extensive insight is given in Section 5.11.

5.3.3.1 Nanocellulose as a Bio-filler

Nanocellulose is a cost-effective component derived from plant sources or microbial synthesis and it could be used as a reinforcing agent in the production of bio-based packaging nanocomposites. Nanocellulose is obtained in the form of nanofibers (CNFs) or crystals (CNCs) *via* combined modifications including chemical, enzymatic or mechanical processes leading to reduced particle size.[2] The good filler capacity and better dispersion in biopolymeric matrices of these nanostructures lead to enhanced functional properties (mechanical, gas and water barrier properties) of the final nanocomposite formulations.[109] Preparation techniques for nanocellulose production have been extensively reviewed and studied.[2,14,22,109] Among the various strategies employed for nanocellulose production, acid hydrolysis with sulfuric acid is the most widely used.[2,109]

Nanocellulose is a highly promising material for packaging applications due to its structural and physicochemical characteristics including structural rigidity, high Young's modulus and tensile strength, low density and high crystallinity with indices between 54–88% as well as simplicity in handling and preparation of films and coatings.[86,109] The incorporation of cellulose nanoparticles in diverse biopolymeric matrices including starch,[114,115] chitosan,[87] gelatin,[116,117] proteins of plant origin[14,86] or combined polymeric blends[118] has been reported.

De *et al.*[114] fabricated bio-based films valorizing agro-industrial wastes. Specifically, waste tamarind seeds were used for starch isolation employed as the biopolymer matrix, while coconut husks were employed for CNC production. The obtained nanocomposite films indicated a significant tensile strength of 21.8 MPa, degradation temperature of 343 °C and moisture absorption capacity of 10.8%. Martelli-Tossi *et al.*[14] prepared SPI-based films reinforced with CNFs or CNCs. According to their results, the nanocomposite films exhibited increased tensile strength by 48% and 38% when 5% of CNFs or CNCs were added to the final formulations, respectively. Young's modulus was also increased by approximately 25.3% and 17% for CNF and CNC additions, respectively, compared to neat SPI-based films. Improved mechanical properties of SPI-based nanocomposites using CNCs have also been reported by Xiao *et al.*[86] and could be attributed to physical chemical interactions of the nanofiller and the polymeric matrix (cross-linking and/or bonding). Increased rigid structure of bio-based nanocomposite films, based on corn starch and chitosan blend, has been associated with increased hydrogen bonds induced by the addition of CNFs.[118]

The type and size of nanocellulose, modification techniques, as well as the concentration are crucial factors affecting the functional properties of the final nanocomposite films. Salari *et al.*[87] demonstrated that the mechanical properties (tensile strength and Young's modulus) of chitosan-based nanocomposite films were positively affected by the addition of bacterial cellulose nanocrystals (BCNs) as a filler up to a 4% concentration. Specifically, increases of 96.1% and 206.4% were observed for tensile strength and Young's modulus, respectively, compared to neat chitosan-based films. However, a further increase in the BCN concentration (6%) resulted in reduced tensile strength and Young's modulus values compared to 4% BCN incorporation.

Similar trends were also observed in the studies of Chaichi *et al.*[89] and Han *et al.*,[119] when various concentrations of CNCs in the range of 2–8% were incorporated in pectin- or SPI-based biofilms, respectively. In the case of pectin-based films, higher tensile strength (13.2 MPa) was observed when 5% CNCs were incorporated, while in the case of SPI-based films, 6% CNCs led to higher tensile strength (11.2 MPa). The reduction in tensile strength after a certain concentration of the nanocellulose filler was added may be attributed to nanocellulose agglomerates leading to uneven nanoparticle distribution in the polymeric matrix.[87,120]

The incorporation of 5% CNFs in gelatin-based nanocomposite films resulted in slightly higher tensile strength (~100 MPa) compared to 5% (w/w) CNC incorporation (~90 MPa), while both CNF- and CNC-reinforced biofilms exhibited similar water and gas barrier behavior.[121]

Modified nanocellulose particles have also been employed as reinforcing fillers to improve the functional properties of the nanocomposite biofilms. Leite *et al.*[116] fabricated gelatin-based nanocomposite films incorporated with CNCs or modified rosin-grafted CNCs. Nanocomposite films reinforced with rosin-grafted CNCs exhibited significantly improved mechanical properties (Young's modulus of 1.9 GPa, tensile strength of 40 MPa) while their water barrier behavior was moderate. The authors reported that the addition of both nanofillers enhanced significantly the tensile strength and Young's modulus of neat-gelatin films, while the incorporation of 4% rosin-grafted CNCs resulted in approximately ~30% higher tensile strength compared to the corresponding nanocomposite films reinforced with CNCs.

Elongation at break of nanocomposite films is associated with the flexibility of the final formulations and is often reduced upon the addition of nanocellulosic fillers.[14,86,120] This reduction could be attributed to limited biopolymer chain mobility caused by nanofillers which decrease the free volume in the matrix of the nanocomposite material.[120] Martelli-Tossi *et al.*,[14] reported decreased elongation at break when CNFs and CNCs were incorporated in SPI-based films compared to neat SPI-based biofilms. However, a combined increase in elongation at break along with tensile strength has also been reported.[89,122] According to Chaichi *et al.*[89] this could be attributed to a good interaction between the biopolymer matrix and the nanocellulose filler leading to successful transferring of the applied stress through the nanocomposite material layers and thus to equal stress distribution and reduced stress concentrations.

Incorporation of nanocellulose as a filler into biopolymeric matrices results in the formation of a tortuous network in which nanocellulose acts as a physical barrier hindering the movement of gases as well as water vapor.[123] Specifically, WVP reduction upon the addition of nanofillers could be attributed to increased crystallinity of final nanocomposite formulations or reduction of hydrophilic groups in the biopolymer matrix leading to the formation of the tortuous network.[116]

The WVP of gelatin-based biofilms was reduced by 55% after incorporation of rosin-grafted CNCs (0.5%).[116] Increasing concentrations of nanocellulose fillers up to specific thresholds have been reported to lead to decreasing trends of WVP as well as gas barrier properties of biofilms based on starch,[115] SPI[86] or biopolymer blends of starch and chitosan.[118] However, the free volume that may be observed in the matrix of the nanocomposite material due to surface incompatibility[123] of the nanocomposite material components or issues concerning reduced homogeneity of nanofiller distribution in the polymeric matrices[14] may lead to inferior gas and water vapor barrier properties.

Yu *et al.*[118] demonstrated that starch/chitosan-based edible packaging films reinforced with a high CNF concentration showed limited microbial growth when they were applied on fresh beef. This could be attributed to the positive surface charges observed in CNFs upon acid fabrication promoting interactions with the negatively charged microbial membrane and thus resulting in serious losses of intracellular components. Additionally, the favorable oxygen barrier properties achieved through the addition of CNFs could limit oxygen availability inside the packaging environment, resulting in aerobic bacteria growth inhibition.

5.4 Functional Biopolymers for Active Food Packaging

Bio-based packaging is gaining extreme interest due to sustainability and biodegradability issues. Product quality and safety assurance combined with prolonged shelf life of edibles, active and edible packaging materials lead to a transition of packaging development to a new upgraded level.[25]

Biodegradable packaging systems in the form of films or coatings have been suggested as alternative packaging materials of edible products. These novel packaging materials have been applied in agricultural or minimally processed fruits and vegetables,[38,84,86,92] meat,[85,86,124,125] poultry[126] and fish[113,127] products, dairy[32,36,116] as well as bakery and confectionary products.[19,88]

5.5 The Case of Proteins

Proteins could be employed for the production of alternative packaging materials due to their excellent film-forming capacity. Their amphiphilic nature, denaturation and electrostatic charges are highly associated with their film-forming capacity and distinguish proteins from other film-forming polymers.[128] The amino acid composition influences the conformational structure of proteins leading to varying functional properties.[129] Thus, specific structural modifications including chemical (acidic or alkaline) or mechanical treatment, cross-linking, heat denaturation and others could result in controlled functional properties of the final bio-based packaging material.[77] Transparency, gas barrier behavior and protein–antioxidant complexes constitute some characteristics of protein-based films that are preferable for packaging applications.[42] However, the poor water barrier properties and brittle nature of the final formulations are major drawbacks limiting their wide application as packaging materials.[10] The native fragile structure of protein-based packaging materials is attributed to high intramolecular disulfide, hydrogen and electrostatic interactions of the protein chains.[77] Plasticizers or/and blending components and fillers could improve all the aforementioned factors.

The functional behavior of protein-based films is highly influenced by the concentration and type (molecular weight, polarity) of the plasticizer. Plasticizers of low molecular weight are preferable, with glycerol being the most commonly used. Kaewprachu *et al.*[80] examined the effect of various plasticizers including glycerol, sorbitol or a combination of both on the final properties of fish myofibrillar protein (FMP)-based films. In binary glycerol/sorbitol systems increasing glycerol concentration led to a significant increase in elongation at break, WVP and oxygen barrier properties, while under increasing sorbitol concentrations enhanced tensile strength (12.6–3.5 MPa) was observed.

Protein chain interactions are highly influenced by pH that further affects their ionization state. pH values higher than the isoelectric point of the corresponding protein favor solubility of proteins and induce disulfide bonds further leading to more stable protein structures and thus superior mechanical properties of the final film formulations.[130] Amado *et al.*[130] produced SPI-based films by a casting method and evaluated the effect of pH on mechanical and gas barrier properties of films. Specifically, decreased tensile strength and elongation at break by 27% and 83% were observed when pH was reduced from 11 to 3, indicating less ordered arrangements. Oxygen permeability was increased to 1.7×10^{15} $g\,m^{-2}\,s^{-1}\,Pa^{-1}$. Alkaline pH conditions up to 9 have been reported to improve the water uptake capacity of gluten-based bioplastics produced *via* extrusion.[76]

The preparation methods employed for protein-based films affect the final properties of the packaging materials. High-intensity ultrasonication could result in aggregation or/and disassociation of protein matrix particles affecting protein chain interactions.[131] Fragmentation of proteins under ultrasonic-assisted treatments[131,132] could disrupt protein intramolecular forces, promoting the availability of non-polar protein fractions for further interaction. Cheng and Cui[131] valorized pea protein isolates to produce edible films and they evaluated the effect of ultrasound-assisted treatment on their functional properties. Ultrasound-assisted preparation of pea protein-based films resulted in increased tensile strength values (1.5–2.2 MPa) combined with increased elongation at break (77.5–88.9%) and enhanced WVP compared to control films. Jiménez-Rosado *et al.*[76] produced gluten-based bioplastics *via* compression molding and extrusion. Compression molding formulations exhibited significantly decreased mechanical properties compared to extrusion-prepared bioplastics.

Physical treatments, *i.e.* ultrasonication, irradiation and thermal treatment could enhance protein cross-linking reactions. During thermal treatment, denaturation and unfolding of proteins occur leading to higher exposure of the hydrophilic groups of proteins to the solvent. As a result, the formation of a more cohesive network is developed due to interchain linkages formed after solvent evaporation.[77]

Cross-linking (chemical or enzymatic) has also been applied as an effective method to overcome the drawbacks associated with functional properties of protein-based films.[77,133] Cross-linking agents provide the potential of a strong interconnection of molecules leading to increased molecular weights and thus induce stability and enhance mechanical and barrier properties.[134] Chemical cross-linking using aldehydes is a common practice. Glutaraldehyde is the predominant cross-linker offering improved mechanical properties of protein-based films and this feature has been associated with the degree and position of cross-linking reactions.[135] Other components such as glyoxal[76] or tannic acid[81] have also been applied as cross-linking agents resulting in improved targeted functional properties. However, major drawbacks of cross-linking are associated with reduced biodegradability, reduced functional group availability and rheology and safety issues arising from the potential cytotoxic effects of cross-linking agents.[94,134]

Enzymatic cross-linking has been suggested as an effective method providing desirable functional properties of protein-based films. Collagen-based films produced *via* enzymatic cross-linking with gallotannin hydrolysates catalyzed by laccase exhibited improved tensile strength (87.2–100.7 MPa) compared to their respective counterparts where only gallotannin hydrolysates were incorporated in the collagen matrix (85.6–95.6 MPa) while improved thermal stability was also observed.[133] Liu *et al.*[94] employed enzymatic cross-linking with transglutaminase to produce SPI–sodium caseinate composites. It was demonstrated that tensile strength and hydrophobicity were significantly improved while elongation at break, solubility in water and WVP were decreased by 71%, 37% and 32%, respectively, compared to control samples.

Blending of proteins with other polymers has also been employed as a common technique for high-performance bio-based packaging materials. Polysaccharides such as pectins[36] and chitosan[24] have been suggested as blending agents with proteins, resulting in the development of an effective 3D network with improved functional properties. Lipids have also been incorporated in protein matrices to enhance water resistance and water vapor barrier properties of protein-based materials that are usually poor due to the hydrophilic nature of proteins.[42] The effect of endothermic polymerization of heat-denatured proteins leads to the formation of protein–lipid composites. In this case lipids interact *via* hydrophobic reactions with proteins and are distributed as droplets in the protein network.[35] Based on the protein : lipid ratio cohesion forces in the composite network may decrease due to lipid molecules promoting fracture phenomena. Incorporation of lipids into protein matrices leads to decreased values of WVP and increased gas barrier properties due to higher solubility of gases in lipidic phases.[77]

The incorporation of functional ingredients into protein-based films or coatings has been applied to develop active packaging materials. Various protein blends or sole proteins including whey protein, gelatin, soy protein among others as well as a wide variety of antioxidant and antimicrobial compounds have been valorized for active food packaging development (Table 5.2).

Polyphenols are commonly incorporated in bio-based formulations to produce active packaging materials as alternatives to synthetic substances. Their binding with proteins is affected by temperature and pH. Greater interactions between proteins and polyphenols have been observed under specific temperatures (*i.e.*, helianthin and caffeic acid under 45 °C) as thermal treatments may promote covalent, ionic or H-bonding. A similar effect may be observed under different pH values leading to the exposure of protein-binding sites and enables ionic interactions with polyphenols.[129]

The addition of antimicrobial or antioxidant substances in biopolymeric formulations leads to controlled release.[77] Reyes *et al.*[137] developed gelatin-based packaging films reinforced with red propolis ethanolic extract combined with clove and/or basil essential oils. The incorporation of combined active components into the gelatin-based films led to increased total

Table 5.2 Recent studies including protein-based films reinforced with various active components.

Source	Reinforcer	Product application	Results	Ref.
Gelatin	MEO 0.38%	—	Effective inhibition of fungal growth	37
Zein/gelatin	Tea polyphenols	Kiwifruit, avocado banana	Effective microbial growth control and color stability of fresh fruits	38
Fish gelatin/pectin	HT/DHPG	Strawberries	Delayed microbial growth and prolonged shelf life	91
Whey/furcellaran	Yerba mate/white tea extracts	Soft cheese	Extended shelf life, lower yeast and mold counts and bacterial growth inhibition	32
Quinoa/chitosan	Thymol NPs 5%	Tomato cherries and blueberries	Inhibition of *Botrytis cinerea* growth, significant reduced weight loss	92
Whey/chitosan	Cranberry or quince juice	Turkey slices	Delayed microbial deterioration for at least 6 days	136
Whey	Oregano essential oils	Kasar cheese	Higher antimicrobial activity against *E. coli* O157:H7	39
	Natamycin		Higher inhibition of *Penicillium* spp. and yeasts and molds	
Gelatin	Cloves and basil essential oils	—	Antimicrobial activity against all tested microorganisms	137
Soy	Diatomite/thymol complex	Blueberries	Delayed deterioration during 5 days of storage	138
Whey	Green tea extract	Salmon	Delayed lipid oxidation until the 14th day of storage	139

phenolic content and increased antioxidant activity and antimicrobial activity against *Staphylococcus aureus* and Salmonella Enteritidis. Brink *et al.*[136] observed increased antimicrobial activity of whey protein/chitosan-based composites supplemented with cranberry or quince juices when they were applied in fresh turkey meat. In the same study, diffusion assays indicated that release and diffusion of juices' active substances from the bio-based films resulted in effective inhibition of pathogenic *Campylobacter jejuni*, *Salmonella agona* and *Salmonella typhimurium* as well as lactic acid bacteria. Delayed lipid oxidation and thus improved shelf life of fresh salmon during storage were achieved when active whey protein-based reinforced with green tea extract was applied.[139]

5.6 Ethylene Scavenging

Ethylene levels in fresh foods must be controlled as this molecule is responsible for fruit and vegetable ripening and senescence. Ethylene scavengers such as potassium permanganate on silica gels or alumina oxidize ethylene and change its color from purple to brown.[140] Dispersed zeolite, active carbon and other minerals which are dispersed in films are often used in packaging for fresh products.[141] They not only absorb ethylene but also change the diffusion properties of the film toward carbon dioxide and oxygen. Ethylene can be oxidized by metals and metal oxides. TiO_2 nanopowder included in polyethylene-based films was used for the packaging of strawberries in refrigeration storage. The sensory and physiological properties of the fruit were preserved in a better condition than in conventional polyethylene bags.[142]

5.7 Antimicrobial Activity

Food spoilage is related to changes in the physical, chemical or biological properties of food that make it unacceptable for consumption.[143] As microorganisms are responsible for most food spoilage, an active food packaging material should reduce access of the microorganisms to the actual food and produce a hostile environment to prevent their population growth.[144] Antimicrobials and other active compounds can be loaded and released from edible films in order to extend the shelf life and ensure the safety of food products. Several materials have been proposed as antimicrobial agents, *i.e.* organic acids, bacteriocins, fungicides, enzymes, polysaccharides, soy sauce, spices, plant extracts and metal NPs.[145] Fruits, vegetables, grain spices and grain extracts contain phenolic compounds that have antimicrobial, anti-inflammatory and antioxidant actions. In addition, essential oils act against foodborne microorganisms and are used in the food industry as preservative and fragrant additives.[146–148]

Chitosan and gelatin are biopolymers with antimicrobial activity. Chitosan's positive amino groups interact with the negatively charged groups of the cell membrane of the pathogenic microorganisms. Similarly, in gelatin the antimicrobial action is related to its amino groups.[149] In order to obtain films with effective antibacterial action these biopolymers have to be supported by adding other active compounds. Metallic and metal oxide NPs are particularly interesting because they generally improve the mechanical and barrier properties of films owing to their large surface area to volume ratio. Moreover, they have antibacterial properties as there is a high contact area available for interaction with the microorganisms.[150]

5.8 Antioxidant Properties

Microorganism growth leads to the oxidation of lipids in meat products. Oxidation has consequences on the flavor and odor of meat and therefore negatively affects the choice of the consumers.[151] In addition, toxic compounds, *e.g.* aldehydes, are formed and the nutritional value of the food is compromised. Therefore, it is important for the meat industry to control oxidation during processing and storage. Active packaging materials may contain active antioxidant ingredients to hinder meat deterioration. Customarily, storing the products under vacuum or in an N_2 atmosphere is applied, however, even small amounts of O_2 can cause oxidation and, on the other hand, an atmosphere rich in O_2 is needed for the preservation of red color.[152] Synthetic antioxidants may effectively protect meat products, but the current consumer trend is to turn to natural antioxidants. Plant extracts and essential oils are promising alternatives which also take advantage of agro-industrial by-products.[153]

5.9 Emission of CO_2

Carbon dioxide has antimicrobial properties that are used in the food industry for food preservation and to increase shelf life.[140] Changes to cytoplasmic pH, inhibition of bacterial enzymes and diffusion and accumulation in the bacterial cell membrane are the main causes of CO_2 antibacterial activity.[154,155] In conventional packaging the head-space gas volume is kept sufficiently high so as to provide adequate dissolution inside the food product and at the same time preserve the shape of the package. Traditionally, the emission of CO_2 is based on the combination of sodium bicarbonate with an organic acid (*e.g.*, citric acid). Liquid from the food triggers their reaction to produce CO_2. Another combination of active ingredients is ascorbic acid with iron carbonate where ascorbic acid antioxidant properties are also utilized.[141] Pads or sachets are conventionally used to carry the active components of the CO_2 emitters. The absorption of food liquids initiates the release of CO_2 in fish, meat and chicken products extending the sensory and microbial shelf life.

Both aqueous and fatty phases are able to dissolve CO_2 which is very important as the amount of dissolved CO_2 defines to a large extent its antimicrobial activity.[141] In food packaging a significant amount of the gas is needed in the headspace for effective antibacterial activity. Therefore, systems that are able to control the atmosphere of the food product releasing CO_2 in a controllable manner are very important. Emitters in packages with modified atmosphere can provide high filling degree, reduced package size and low environmental burden. In meat storage with free CO_2, collapse of the packages occurs upon dilution of CO_2 in the meat and drip loss increases, which are both considered undesirable effects by customers.[156] A CO_2-emitting sachet or pad compensates for the CO_2 that dissolves in the food and it absorbs the water- and sarcoplasmic protein-containing drip. The sodium bicarbonate and citric acid react as they dissolve and CO_2 is emitted. The equilibrium will be directed toward the production of water and CO_2 in the acidic environment created by the organic acid where carbonic acid is neutralized.

5.10 Scavenging of O_2

Films with oxygen scavenging provide extended shelf life, preserve nutrition quality, inhibit discoloration and ensure safety of foods.[157] The commonly used vacuum technology and modified atmosphere packaging deplete oxygen from the headspace up to a certain level, however, some oxygen can be trapped and dispersed in the food.[158] Solution casting was used to prepare chitosan films with gallic acid and sodium carbonate with active oxygen barrier properties.[159] Gallic acid affected the mechanical and structural characteristics of the films. Films with gallic acid and sodium carbonate had higher water solubility and thickness in comparison with films of neat chitosan. Gallic acid resulted in lower oxygen and water permeability. Oxygen scavenging was evaluated in sealed glass flasks containing the films and an O_2 sensor sticker. The O_2 scavenging was more effective as the gallic acid concentration increased. In nanocomposite films of poly(ethylene-*co*-vinyl alcohol), cellulose nanocrystals were used to achieve good dispersion of palladium NPs leading to enhanced oxygen scavenging.[160]

5.11 Organic–Inorganic Hybrid Biopolymers in Food Packaging

Metallic NPs are very broadly used as inorganic inclusions in food packaging. Au NPs were synthesized *in situ* by adding $HAuCl_4$ and tri-sodium citrate to starch aqueous solutions. These starch/Au bionanocomposites were investigated for their potential as packaging materials. The size of the spherical NPs was 20–30 nm. The antimicrobial action of the bionanocomposites against four bacteria and three fungi was enhanced by an increase in the Au NP amount.[161] Quinoa starch solutions mixed with Au NP dispersions were used to prepare casted biofilms. These films showed high antibacterial activity against pathogens born in foods.[162] Drop-casted Au NP-containing films based on chitosan were obtained by the previous formation of the NPs (4 nm) in chitosan solutions. Aminopropylsilane was added to the dispersions after the preparation of the NPs. The three components acted synergistically against the *Salmonella* bacterium.[163] Alginate hydrogels cross-linked by calcium chloride and *in situ* synthesized NPs were fabricated as thermal history indicators for food packaging. The localized surface plasmon resonance of the AuNPs was the physical property that was exploited. The color of the hydrogel films changed from gray to red upon treatment with high temperature. Moreover, the time and temperature (thermal history) dependence of the color change could be tuned by the composition of the hybrid material. The hydrogels were proposed as reliable and inexpensive active packaging for perishable products.[164]

Agar films doped with silver NPs were designed for active packaging. Mechanical strength, water vapor barrier properties and surface hydrophobicity could also be enhanced by the increase in the amount of Ag NPs. Their activity against bacterial pathogens was significant above a 1% content of NPs. The plasmonic peak was broadened and shifted to higher wavelengths at these concentrations because of agglomeration between the NPs. XRD revealed the broad peaks of alginate at low angles and the sharp diffraction peaks of crystalline Ag at higher angles.[165] Polyethylene films were coated by chitosan solutions that incorporated AgNPs and laurel essential oil-loaded liposomes. The release of the metallic NPs and the essential oil were responsible for the antimicrobial and antioxidant properties of the hybrid films. Antimicrobial activity was tested by the plate method on *S. aureus* and *E. coli*. The scavenging capacity of the liposomes against DPPH proved that they were very efficient for the elimination of free radicals. Storage studies of pork meat showed that polyethylene films preserve food for up to 9 days, films coated only with chitosan prolonged the shelf life for 3 more days, while the incorporation of the loaded liposomes prolonged preservation for another 3 days.[166]

ZnO–Ag NPs 8–40 nm in size were grown on chitin nanowhiskers (150–260 nm in length) using zinc acetate dihydrate mixed with chitin in suspension. The suspended NP-loaded whiskers were mixed with carboxymethyl cellulose to prepare nanocomposite films. The viable cell colony count method on *L. monocytogenes* or *E. coli* showed that the composite material had very weak antimicrobial activity on *E. coli* and none on *L. monocytogenes* when no NPs were incorporated. In that case the interaction between the positively charged chitin and the cell membrane of the bacteria led to cell death. When ZnO–Ag NPs were included strong bacterial activity was observed which was higher against *E. coli*.[167] In another study of biopolymer films based on carboxymethylcellulose, ZnO NPs and plant-sourced cinnamaldehyde were used as inclusions. The films had good mechanical, optical and water barrier properties. Disc inhibition zone assays were applied to evaluate the antifungal activity of the hybrid films against *A. niger*. Strong inhibition was observed when either cinnamaldehyde or ZnO NPs were included. A synergistic effect of the two components was found as their combined effect was the strongest. It was proposed that the moderate oxidative stress response produced by ZnO NPs and cinnamaldehyde individually added up to very intense ROS accumulation in the fungus. The films were found to improve the postharvest storage quality of cherry tomatoes.[168]

The shelf life of packaged red grapes was greatly improved by chitosan films that were doped with titanium dioxide nanopowder. The films showed enhanced mechanical properties, higher hydrophilicity and good antimicrobial activity in visible light. Interestingly, it was found that the chitosan–TiO_2 films caused death of the cells by changing the structure of the membrane and causing leakage of intracellular substances. Red grapes wrapped with a plastic bag (control) changed to dark red with a mildew appearance and emitted a rotten odor after 6 days preservation at 37 °C. Grapes stored with the bionanocomposite films retained their juice in their interior and were bright. Mildew did not occur for 15 days for chitosan films and for 22 days for the composite chitosan–TiO_2 films.[169] Nanocomposite packaging material from whey protein isolate, cellulose nanofibers, TiO_2 and rosemary essential oil were investigated for the preservation of lamb meat under refrigeration. The shelf life increased and the microbial and sensory qualities of the lamb meat were preserved. In more detail, a sensory assessment was performed where color, odor, texture and overall satisfaction were surveyed. The whole set of sensory attributes was satisfactory for both the control (plastic bag) and treated (nanocomposite) samples at the beginning of storage. Observations every 3 days of storage showed that the sensory attributes of the treated samples were much better preserved in comparison to the control.[170]

Inclusion of SiO_2 NPs into poly(3-hydroxybutyrate-*co*-3-hydroxyhexanoate) films significantly increased the crystallinity of the latter as SiO_2 NPs acted as nucleation sites. In addition, the thermal properties of the films were enhanced as in the one-stage decomposition of poly(3-hydroxybutyrate-*co*-3-hydroxyhexanoate), the end temperature shifted to higher values and the temperature range of the transition increased. The water vapor and oxygen permeability decreased while the films became more hydrophilic by the addition of SiO_2 NPs.[171] Semi-refined iota-carrageenan hybrid films for food packaging were enriched with SiO_2 and ZnO NPs. The degradation of the films was tested in soil. The NP-containing films showed better degradation than the pure iota-carrageenan films. The enzymatic degradation of the polysaccharide chains was caused by the microorganisms in the soil. The combined action of the two kinds of NPs resulted in impedance of biodegradation. It was assumed that either the NPs act as catalysts on the biodegradation and hydrolysis reactions or they interfere with the diffusion and penetration of the living organisms into the films.[172] Al_2O_3 NPs embedded in carboxymethyl cellulose matrices were decorated by SiO_2 and TiO_2 NPs forming core–shell structures. The composite NPs presented low aggregation and good dispersion.[173]

Food preservation was achieved with corn starch-based films that contained halloysite nanoclay and nisin or pediocin. The presence of halloysite nanofiller provided antimicrobial activity and mechanical reinforcement to the films. The bacteriocins adsorbed on the surface of halloysite nanotubes and the nanoclay were responsible for the antimicrobial characteristics of the films. Nisin-containing films were very effective against *L. monocytogenes*.

In the case of pediocin inhibition halos were observed for *L. monocytogenes*, while the inhibition area was restricted to the periphery of the films for *C. perfringens*.[174] Nanocomposite films with halloysite have been also prepared by solution casting using solutions containing chitosan and starch. The interaction between the nanoclay and the biopolymers was proven by FTIR and XRD, while the films were found to be hydrophilic as confirmed by contact angle measurements. These films were proposed for wound healing purposes, however, their properties also are attractive for food packaging.[175] Halloysite nanotubes have been functionalized with NaOH and ZnO NPs and incorporated into alginate film. The reinforcement led to enhancement of the mechanical and water and UV-light barrier properties. These films were also effective in sterilization against foodborne pathogenic bacteria.[176]

Films of cassava starch with cinnamon essential oil were reinforced by clay bentonite NPs. The films had strong antibacterial action against *Escherichia coli*, *Salmonella typhimurium* and *Staphylococcus aureus*. This effect was an increasing function of the essential oil content. The films were proposed for the storage of meatballs where microbial growth was kept below the FDA limit for 96 h. The physical and mechanical properties of the films were affected by the incorporation of essential oil. The increase in essential oil increased the water vapor permeability of the film despite its hydrophobic nature. However, the tensile strength decreased with the increase of essential oil but increased with the addition of the nanoclay.[177] In another work, organonanoclay was incorporated in polycaprolactone-based films that also contained chitosan. The antimicrobial properties of the films originated mainly from chitosan although the nanoclay showed antimicrobial action as well. The mechanical properties of the films were weaker as more chitosan was added, however, increasing the amount of chitosan resulted in shorter degradation times.[178] Natural orange peel extract and Cloisite were used as additives in biodegradable and antibacterial protein isolate/gelatin nanocomposite films. Orange peel extract content increased mechanical properties, water solubility and antibacterial activity. The incorporation of dispersed clay led to films of increased thickness.[179]

5.12 Future Perspectives and Challenges

Rising consumer awareness of environmental issues combined with unstable petroleum prices have shifted scientific interest toward alternative materials that are eco-friendly and do not compete with the food chain. Biopolymers are potential candidates for the substitution of fossil-based resources due to properties including versatility, biodegradability, renewability and sustainability. The large-scale production of biopolymers derived from renewable resources is still limited due to the low availability, seasonality and fractionation of the latter. This is interlinked with high prices and thus deteriorated applications. Optimal functional properties of bio-based packaging materials are challenging to promote industrial cost-effective applications and thus investigation of the film-forming mechanisms is critical.

Biodegradable packaging films and coatings including polysaccharide-, protein- and lipid-based films of plant, animal and microbial origin are of enormous interest in the food industry sector, and should comply with norms of quality safety and shelf life and preservation of food products. Biopolymers produced *via* microbial fermentation, *i.e.* PHAs and BC, as well as renewable proteins of high purity derived from waste biomass, *i.e.* sunflower meal, are under intense investigation. Issues of poor mechanical and water or gas barrier properties could be overcome by employing plasticizing or filling agents, various physical, chemical or enzymatic modifications as well as hybrid and blending formulations. Nanotechnology could further improve their performance and multifunctionality using nanoparticle fillers as reinforcing agents. Additionally, these nanomaterials could be used as active carriers of antioxidants and antimicrobials, positively affecting shelf life and nutritional quality and of highly perishable products during storage. Active packaging is a novel technology while several factors need to be considered for successful formulation and application including compatibility and synergistic interactions of components and food physicochemical characteristics. The preparation method of biofilms, *i.e.* solvent casting, compression molding and extrusion, is also a critical factor in the production of competitive final formulations with similar physicochemical properties to their conventional counterparts.

References

1. P. Cazón and M. Vázquez, Bacterial cellulose as a biodegradable food packaging material: A review, *Food Hydrocolloids*, 2021, **113**, 106530.
2. D. Selianitis *et al.*, Nanocellulose production from different sources and their self-assembly in composite materials, in *Handbook of Nanocelluloses*, ed. A. Barhoum, Springer International Publishing, 2021.
3. P. Cazón, G. Velazquez, J. A. Ramírez and M. Vázquez, Polysaccharide-based films and coatings for food packaging: A review, *Food Hydrocolloids*, 2017, **68**, 136–148.
4. H. Wang, J. Qian and F. Ding, Emerging Chitosan-Based Films for Food Packaging Applications, *J. Agric. Food Chem.*, 2018, **66**, 395–413.
5. M. K. Lauer and R. C. Smith, Recent advances in starch-based films toward food packaging applications: Physicochemical, mechanical, and functional properties, *Compr. Rev. Food Sci. Food Saf.*, 2020, **19**, 3031–3083.
6. A. A. Oun and J.-W. Rhim, Carrageenan-based hydrogels and films: Effect of ZnO and CuO nanoparticles on the physical, mechanical, and antimicrobial properties, *Food Hydrocolloids*, 2017, **67**, 45–53.
7. L. L. Hyland, M. B. Taraban and Y. B. Yu, Using Small-Angle Scattering Techniques to Understand Mechanical Properties of Biopolymer-Based Biomaterials, *Soft Matter*, 2013, **9**, 10218–10228.

8. M. F. Moradali and B. H. A. Rehm, Bacterial biopolymers: from pathogenesis to advanced materials, *Nat. Rev. Microbiol.*, 2020, **18**, 195–210.
9. E. M. N. Polman, G.-J. M. Gruter, J. R. Parsons and A. Tietema, Comparison of the aerobic biodegradation of biopolymers and the corresponding bioplastics: A review, *Sci. Total Environ*, 2021, **753**, 141953.
10. M. Kurek and N. Benbettaieb, Bio-based Materials, *Packaging Materials and Processing for Food, Pharmaceuticals and Cosmetics*, John Wiley & Sons, Ltd, 2021, pp. 145–182.
11. H. Haghighi, *et al.*, Comparative analysis of blend and bilayer films based on chitosan and gelatin enriched with LAE (lauroyl arginate ethyl) with antimicrobial activity for food packaging applications, *Food Packag. Shelf Life*, 2019, **19**, 31–39.
12. A. A. Tyuftin and J. P. Kerry, Gelatin films: Study review of barrier properties and implications for future studies employing biopolymer films, *Food Packag. Shelf Life*, 2021, **29**, 100688.
13. F. Pavli, *et al.*, Antimicrobial Activity of Oregano Essential Oil Incorporated in Sodium Alginate Edible Films: Control of Listeria monocytogenes and Spoilage in Ham Slices Treated with High Pressure Processing, *Materials*, 2019, **12**, 3726.
14. M. Martelli-Tosi, *et al.*, Soybean straw nanocellulose produced by enzymatic or acid treatment as a reinforcing filler in soy protein isolate films, *Carbohydr. Polym.*, 2018, **198**, 61–68.
15. X. Zhou, *et al.*, Biodegradable sandwich-architectured films derived from pea starch and polylactic acid with enhanced shelf-life for fruit preservation, *Carbohydr. Polym.*, 2021, **251**, 117117.
16. Q. Xu, C. Chen, K. Rosswurm, T. Yao and S. Janaswamy, A facile route to prepare cellulose-based films, *Carbohydr. Polym.*, 2016, **149**, 274–281.
17. Z. Yu, Y. Ji, V. Bourg, M. Bilgen and J. C. Meredith, Chitin- and cellulose-based sustainable barrier materials: a review, *Emergent Mater*, 2020, **3**, 919–936.
18. R. Priyadarshi and J.-W. Rhim, Chitosan-based biodegradable functional films for food packaging applications, *Innovative Food Sci. Emerging Technol.*, 2020, **62**, 102346.
19. N. Noshirvani, B. Ghanbarzadeh, R. Rezaei Mokarram and M. Hashemi, Novel active packaging based on carboxymethyl cellulose-chitosan-ZnO NPs nanocomposite for increasing the shelf life of bread, *Food Packag. Shelf Life*, 2017, **11**, 106–114.
20. A. H. Bedane, M. Eić, M. Farmahini-Farahani and H. Xiao, Water vapor transport properties of regenerated cellulose and nanofibrillated cellulose films, *J. Membr. Sci.*, 2015, **493**, 46–57.
21. S. Sivakanthan, S. Rajendran, A. Gamage, T. Madhujith and S. Mani, Antioxidant and antimicrobial applications of biopolymers: A review, *Food Res. Int.*, 2020, **136**, 109327.
22. H. Du, *et al.*, Cellulose nanocrystals and cellulose nanofibrils based hydrogels for biomedical applications, *Carbohydr. Polym.*, 2019, **209**, 130–144.

23. Cellulose Market Size, Share, Growth Industry Report 2026. Fortune business insights, https://www.fortunebusinessinsights.com/cellulose-market-102062.
24. L. Zhang, *et al.*, The properties of chitosan/zein blend film and effect of film on quality of mushroom (Agaricus bisporus), *Postharvest Biol. Technol.*, 2019, **155**, 47–56.
25. J. Huang, *et al.*, Pectin-based active packaging: A critical review on preparation, physical properties and novel application in food preservation, *Trends Food Sci. Technol.*, 2021, **118**, 167–178.
26. M. H. Shafie, R. Yusof, D. Samsudin and C.-Y. Gan, Averrhoa bilimbi pectin-based edible films: Effects of the linearity and branching of the pectin on the physicochemical, mechanical, and barrier properties of the films, *Int. J. Biol. Macromol.*, 2020, **163**, 1276–1282.
27. A. Rehman, *et al.*, Pectin polymers as wall materials for the nano-encapsulation of bioactive compounds, *Trends Food Sci. Technol.*, 2019, **90**, 35–46.
28. J. Meerasri and R. Sothornvit, Characterization of bioactive film from pectin incorporated with gamma-aminobutyric acid, *Int. J. Biol. Macromol.*, 2020, **147**, 1285–1293.
29. U. Amin, *et al.*, Potentials of polysaccharides, lipids and proteins in biodegradable food packaging applications, *Int. J. Biol. Macromol.*, 2021, **183**, 2184–2198.
30. Y. Fan, J. Yang, A. Duan and X. Li, Pectin/sodium alginate/xanthan gum edible composite films as the fresh-cut package, *Int. J. Biol. Macromol.*, 2021, **181**, 1003–1009.
31. X. Huang, *et al.*, Formation mechanism of egg white protein/κ-Carrageenan composite film and its application to oil packaging, *Food Hydrocolloids*, 2020, **105**, 105780.
32. A. Pluta-Kubica, E. Jamróz, A. Kawecka, L. Juszczak and P. Krzyściak, Active edible furcellaran/whey protein films with yerba mate and white tea extracts: Preparation, characterization and its application to fresh soft rennet-curd cheese, *Int. J. Biol. Macromol.*, 2020, **155**, 1307–1316.
33. S. A. A. Mohamed, M. El-Sakhawy and M. A.-M. El-Sakhawy, Polysaccharides, protein and lipid-based natural edible films in food packaging: A Review, *Carbohydr. Polym.*, 2020, **238**, 116178.
34. A. Poirier, *et al.*, Sunflower Proteins at Air–Water and Oil–Water Interfaces, *Langmuir*, 2021, **37**, 2714–2727.
35. V. N. Enujiugha and A. M. Oyinloye, Protein-Lipid Interactions and the Formation of Edible Films and Coatings, in *Encyclopedia of Food Chemistry*, ed. L. Melton, F. Shahidi and P. Varelis, Academic Press, 2019. pp. 478–482.
36. M. Jridi, *et al.*, Physicochemical, antioxidant and antibacterial properties of fish gelatin-based edible films enriched with orange peel pectin: Wrapping application, *Food Hydrocolloids*, 2020, **103**, 105688.

37. L. Scartazzini, *et al.*, Gelatin edible coatings with mint essential oil (Mentha arvensis): film characterization and antifungal properties, *J. Food Sci. Technol.*, 2019, **56**, 4045–4056.
38. C. Xia, W. Wang, L. Wang, H. Liu and J. Xiao, Multilayer zein/gelatin films with tunable water barrier property and prolonged antioxidant activity, *Food Packag. Shelf Life*, 2019, **19**, 76–85.
39. A. C. Seydim, G. Sarikus-Tutal and E. Sogut, Effect of whey protein edible films containing plant essential oils on microbial inactivation of sliced Kasar cheese, *Food Packag. Shelf Life*, 2020, **26**, 100567.
40. J. C. P. Santos, *et al.*, Nisin and other antimicrobial peptides: Production, mechanisms of action, and application in active food packaging, *Innovative Food Sci. Emerging Technol.*, 2018, **48**, 179–194.
41. S. Gopalakrishnan, J. Xu, F. Zhong and V. M. Rotello, Strategies for Fabricating Protein Films for Biomaterial Applications, *Adv. Sustainable Syst.*, 2021, **5**, 2000167.
42. H. Chen, *et al.*, Application of Protein-Based Films and Coatings for Food Packaging: A Review, *Polymers*, 2019, **11**, 2039.
43. I. Assad, S. U. Bhat, A. Gani and A. Shah, Protein based packaging of plant origin: Fabrication, properties, recent advances and future perspectives, *Int. J. Biol. Macromol.*, 2020, **164**, 707–716.
44. J. Gómez-Estaca, R. Gavara, R. Catalá and P. Hernández-Muñoz, The potential of proteins for producing food packaging materials: a review, *Packag. Technol. Sci.*, 2016, **29**, 203–224.
45. USDA, United States Department of Agriculture (Foreign Agricultural Service), https://apps.fas.usda.gov/psdonline/app/index.html#/app/advQuery.
46. M.-N. Efthymiou, C. Pateraki, H. Papapostolou, C. S. K. Lin and A. Koutinas, Restructuring the sunflower-based biodiesel industry into a circular bio-economy business model converting sunflower meal and crude glycerol into succinic acid and value-added co-products, *Biomass Bioenergy*, 2021, **155**, 106265.
47. P. R. Salgado, M. E. López-Caballero, M. C. Gómez-Guillén, A. N. Mauri and M. P. Montero, Sunflower protein films incorporated with clove essential oil have potential application for the preservation of fish patties, *Food Hydrocolloids*, 2013, **33**, 74–84.
48. P. R. Salgado, S. E. Molina Ortiz, S. Petruccelli and A. N. Mauri, Biodegradable sunflower protein films naturally activated with antioxidant compounds, *Food Hydrocolloids*, 2010, **24**, 525–533.
49. Y. Zhang, B. K. Simpson and M.-J. Dumont, Effect of beeswax and carnauba wax addition on properties of gelatin films: A comparative study, *Food Biosci.*, 2018, **26**, 88–95.
50. N. Vishnu Priya, U. G. Vinitha and M. Meenakshi Sundaram, Preparation of chitosan-based antimicrobial active food packaging film incorporated with Plectranthus amboinicus essential oil, Biocatal, *Agric. Biotechnol.*, 2021, **34**, 102021.

51. Y. Jiang, *et al.*, Preparation and characterization of grass carp collagen-chitosan-lemon essential oil composite films for application as food packaging, *Int. J. Biol. Macromol.*, 2020, **160**, 340–351.
52. P. Rai, S. Mehrotra, S. Priya, E. Gnansounou and S. K. Sharma, Recent advances in the sustainable design and applications of biodegradable polymers, *Bioresour. Technol.*, 2021, **325**, 124739.
53. Y. Hu, W. A. Daoud, K. K. L. Cheuk and C. S. K. Lin, Newly Developed Techniques on Polycondensation, Ring-Opening Polymerization and Polymer Modification: Focus on Poly(Lactic Acid), *Materials*, 2016, **9**, 133.
54. K. Hamad, M. Kaseem, M. Ayyoob, J. Joo and F. Deri, Polylactic acid blends: The future of green, light and tough, *Prog. Polym. Sci.*, 2018, **85**, 83–127.
55. N. Mohamad, *et al.*, Development of active agents filled polylactic acid films for food packaging application, *Int. J. Biol. Macromol.*, 2020, **163**, 1451–1457.
56. I. Gan and W. S. Chow, Antimicrobial poly(lactic acid)/cellulose bio-nanocomposite for food packaging application: A review, *Food Packag. Shelf Life*, 2018, **17**, 150–161.
57. M. Asgher, S. A. Qamar, M. Bilal and H. M. N. Iqbal, Bio-based active food packaging materials: Sustainable alternative to conventional petrochemical-based packaging materials, *Food Res. Int.*, 2020, **137**, 109625.
58. M. Alexandri *et al.*, Valorization of fruit processing by-product streams into integrated biorefinery concepts: extraction of value-added compounds and bioconversion to chemicals, in *Valorization of Agri-Food Wastes By-Products*, ed. R. Bhat, Academic Press, 2021, ch. 46, pp. 927–945.
59. Y. Zhong, P. Godwin, Y. Jin and H. Xiao, Biodegradable polymers and green-based antimicrobial packaging materials: A mini-review, *Adv. Ind. Eng. Polym. Res.*, 2020, **3**, 27–35.
60. D. M. Panaitescu, *et al.*, Role of bacterial cellulose and poly (3-hydroxyhexanoate-co-3-hydroxyoctanoate) in poly (3-hydroxybutyrate) blends and composites, *Cellulose*, 2018, **25**, 5569–5591.
61. I. T. Seoane, *et al.*, Effect of Cellulose Nanocrystals and Bacterial Cellulose on Disintegrability in Composting Conditions of Plasticized PHB Nanocomposites, *Polymers*, 2017, **9**, 561.
62. C. Villegas, *et al.*, Obtaining Active Polylactide (PLA) and Polyhydroxybutyrate (PHB) Blends Based Bionanocomposites Modified with Graphene Oxide and Supercritical Carbon Dioxide (scCO2)-Assisted Cinnamaldehyde: Effect on Thermal-Mechanical, Disintegration and Mass Transport Properties, *Polymers*, 2021, **13**, 3968.
63. V. Kachrimanidou, *et al.*, Techno-economic evaluation and life-cycle assessment of poly(3-hydroxybutyrate) production within a biorefinery concept using sunflower-based biodiesel industry by-products, *Bioresour. Technol.*, 2021, **326**, 124711.

64. R. Du, F. Zhao, Q. Peng, Z. Zhou and Y. Han, Production and characterization of bacterial cellulose produced by Gluconacetobacter xylinus isolated from Chinese persimmon vinegar, *Carbohydr. Polym.*, 2018, **194**, 200–207.
65. R. T. A. Machado, *et al.*, Komagataeibacter rhaeticus grown in sugarcane molasses-supplemented culture medium as a strategy for enhancing bacterial cellulose production, *Ind. Crops Prod.*, 2018, **122**, 637–646.
66. L. Ji, F. Zhang, L. Zhu and J. Jiang, An in-situ fabrication of bamboo bacterial cellulose/sodium alginate nanocomposite hydrogels as carrier materials for controlled protein drug delivery, *Int. J. Biol. Macromol.*, 2021, **170**, 459–468.
67. H. M. C. Azeredo, H. Barud, C. S. Farinas, V. M. Vasconcellos and A. M. Claro, Bacterial Cellulose as a Raw Material for Food and Food Packaging Applications, *Front. Sustainable Food Syst.*, 2019, **3**, 7.
68. Q. Yang, *et al.*, Improved thermal and mechanical properties of bacterial cellulose with the introduction of collagen, *Cellulose*, 2017, **24**, 3777–3787.
69. S. Wang, *et al.*, Super-Strong, Super-Stiff Macrofibers with Aligned, Long Bacterial Cellulose Nanofibers, *Adv. Mater.*, 2017, **29**, 1702498.
70. Z. Liu, D. Lin, P. Lopez-Sanchez and X. Yang, Characterizations of bacterial cellulose nanofibers reinforced edible films based on konjac glucomannan, *Int. J. Biol. Macromol.*, 2020, **145**, 634–645.
71. P. Cazón and M. Vázquez, Improving bacterial cellulose films by ex-situ and in-situ modifications: A review, *Food Hydrocolloids*, 2021, **113**, 106514.
72. Plastic Packaging Market Size to Be Worth Around US$586.2 Billion by 2030, Covering the Printing Inks, Coatings and Allied Industries – Ink World, 2021, https://www.inkworldmagazine.com/contents/view_breaking-news/2021-08-20/plastic-packaging-market-size-to-be-worth-around-us5862-billion-by-2030/.
73. Food Packaging Market Size, Share, Growth Report, 2030, Market research future, 2021, https://www.marketresearchfuture.com/reports/food-packaging-market-2086.
74. Global Biodegradable Packaging Solutions Market 2021–26 Industry Share, Size, Growth – Mordor Intelligence, https://www.mordorintelligence.com/industry-reports/gobal-biodegradable-packaging-solutions-market-industry.
75. C. I. La Fuente Arias, M. T. K. Kubo, C. C. Tadini and P. E. D. Augusto, Bio-based multilayer films: A review of the principal methods of production and challenges, *Crit. Rev. Food Sci. Nutr.*, 2021, 1–17.
76. M. Jiménez-Rosado, *et al.*, Bioplastics based on wheat gluten processed by extrusion, *J. Clean. Prod.*, 2019, **239**, 117994.
77. A. Chiralt, C. González-Martínez, M. Vargas and L. Atarés, Edible films and coatings from proteins, in *Proteins Food Process*, ed. R. Y. Yada, Woodhead Publishing, 2nd edn, 2018, ch. 18, pp. 477–500.

78. S. J. A. Rowat, R. L. Legge and C. Moresoli, Plant protein in material extrusion 3D printing: Formation, plasticization, prospects, and challenges, *J. Food Eng.*, 2021, **308**, 110623.
79. F. Zhu, Polysaccharide based films and coatings for food packaging: Effect of added polyphenols, *Food Chem.*, 2021, **359**, 129871.
80. P. Kaewprachu, K. Osako, N. Rungraeng and S. Rawdkuen, Properties of fish myofibrillar protein film: effect of glycerol-sorbitol combinations, *J. Food Sci. Technol.*, 2022, **59**, 1619–1628.
81. M. L. Picchio, *et al.*, Casein films crosslinked by tannic acid for food packaging applications, *Food Hydrocolloids*, 2018, **84**, 424–434.
82. A. González, G. N. Barrera, P. I. Galimberti, P. D. Ribotta and C. I Alvarez Igarzabal, Development of edible films prepared by soy protein and the galactomannan fraction extracted from Gleditsia triacanthos (Fabaceae) seed, *Food Hydrocolloids*, 2019, **97**, 105227.
83. N. L. Vanden Braber, *et al.*, Antifungal whey protein films activated with low quantities of water soluble chitosan, *Food Hydrocolloids*, 2021, **110**, 106156.
84. S. Kamthai and R. Magaraphan, Development of an active polylactic acid (PLA) packaging film by adding bleached bagasse carboxymethyl cellulose (CMCB) for mango storage life extension, *Packag. Technol. Sci.*, 2019, **32**, 103–116.
85. C. Wang, *et al.*, Biopolymer films based on chitosan/potato protein/linseed oil/ZnO NPs to maintain the storage quality of raw meat, *Food Chem.*, 2020, **332**, 127375.
86. Y. Xiao, Y. Liu, S. Kang and H. Xu, Insight into the formation mechanism of soy protein isolate films improved by cellulose nanocrystals, *Food Chem.*, 2021, **359**, 129971.
87. M. Salari, M. Sowti Khiabani, R. Rezaei Mokarram, B. Ghanbarzadeh and H. Samadi Kafil, Development and evaluation of chitosan based active nanocomposite films containing bacterial cellulose nanocrystals and silver nanoparticles, *Food Hydrocolloids*, 2018, **84**, 414–423.
88. Y.-T. Shih and Y. Zhao, Development, characterization and validation of starch based biocomposite films reinforced by cellulose nanofiber as edible muffin liner, *Food Packag. Shelf Life*, 2021, **28**, 100655.
89. M. Chaichi, M. Hashemi, F. Badii and A. Mohammadi, Preparation and characterization of a novel bionanocomposite edible film based on pectin and crystalline nanocellulose, *Carbohydr. Polym.*, 2017, **157**, 167–175.
90. L. Jiang, *et al.*, Development and characterization of zein edible films incorporated with catechin/β-cyclodextrin inclusion complex nanoparticles, *Carbohydr. Polym.*, 2021, **261**, 117877.
91. A. Bermúdez-Oria, G. Rodríguez-Gutiérrez, B. Vioque, F. Rubio-Senent and J. Fernández-Bolaños, Physical and functional properties of pectin-fish gelatin films containing the olive phenols hydroxytyrosol and 3,4-dihydroxyphenylglycol, *Carbohydr. Polym.*, 2017, **178**, 368–377.

92. E. Medina, *et al.*, Chitosan thymol nanoparticles improve the antimicrobial effect and the water vapour barrier of chitosan-quinoa protein films, *J. Food Eng.*, 2019, **240**, 191–198.
93. T. Nisar, *et al.*, Characterization of citrus pectin films integrated with clove bud essential oil: Physical, thermal, barrier, antioxidant and antibacterial properties, *Int. J. Biol. Macromol.*, 2018, **106**, 670–680.
94. K. Liu, *et al.*, Mechanical properties and water sensitivity of soybean protein isolate film improved by incorporation of sodium caseinate and transglutaminase, *Prog. Org. Coat.*, 2021, **153**, 106154.
95. R. Balasubramanian, S. S. Kim and J. Lee, Novel synergistic transparent k-Carrageenan/Xanthan gum/Gellan gum hydrogel film: Mechanical, thermal and water barrier properties, *Int. J. Biol. Macromol.*, 2018, **118**, 561–568.
96. B. R. B. Lara, *et al.*, Water sorption thermodynamic behavior of whey protein isolate/polyvinyl alcohol blends for food packaging, *Food Hydrocolloids*, 2020, **103**, 105710.
97. E. Olsson, M. S. Hedenqvist, C. Johansson and L. Järnström, Influence of citric acid and curing on moisture sorption, diffusion and permeability of starch films, *Carbohydr. Polym.*, 2013, **94**, 765–772.
98. C. Menzel, Improvement of starch films for food packaging through a three-principle approach: Antioxidants, cross-linking and reinforcement, *Carbohydr. Polym.*, 2020, **250**, 116828.
99. G. A. Valencia, C. G. Luciano, R. V. Lourenço, A. M. Q. B. Bittante and P. J. do Amaral Sobral, Morphological and physical properties of nano-biocomposite films based on collagen loaded with laponite®, *Food Packag. Shelf Life*, 2019, **19**, 24–30.
100. Z. Kassab, F. Aziz, H. Hannache, H. Ben Youcef and M. El Achaby, Improved mechanical properties of k-carrageenan-based nanocomposite films reinforced with cellulose nanocrystals, *Int. J. Biol. Macromol.*, 2019, **123**, 1248–1256.
101. W. Zhou, *et al.*, Carboxymethyl chitosan-pullulan edible films enriched with galangal essential oil: Characterization and application in mango preservation, *Carbohydr. Polym.*, 2021, **256**, 117579.
102. A. Papagiannopoulos, Small-Angle Neutron Scattering (SANS), in *Microscopy Methods in Nanomaterials Characterization*, ed. S. Thomas, R. Thomas, A. K. Zachariah and R. K. Mishra, Elsevier, 2017, ch. 10, pp. 339–361.
103. Z. Wu and X. Xing, Small Angle X-ray Scattering and Its Applications, *Synchrotron Radiation Applications*, World Scientific, 2017, pp. 225–285.
104. D. Qiao, *et al.*, Microstructure and Mechanical/Hydrophilic Features of Agar-Based Films Incorporated with Konjac Glucomannan, *Polymers*, 2019, **11**, 1952.
105. V. P. Romani, A. V. Machado, B. D. Olsen and V. G. Martins, Effects of pH modification in proteins from fish (Whitemouth croaker) and their

application in food packaging films, *Food Hydrocolloids*, 2018, **74**, 307–314.

106. I. Benito-González, A. López-Rubio and M. Martínez-Sanz, Potential of lignocellulosic fractions from Posidonia oceanica to improve barrier and mechanical properties of bio-based packaging materials, *Int. J. Biol. Macromol.*, 2018, **118**, 542–551.
107. F. A. Corsello, *et al.*, Morphology and properties of neutralized chitosan-cellulose nanocrystals biocomposite films, *Carbohydr. Polym.*, 2017, **156**, 452–459.
108. M. J. Fabra, M. Martínez-Sanz, L. G. Gómez-Mascaraque, R. Gavara and A. López-Rubio, Structural and physicochemical characterization of thermoplastic corn starch films containing microalgae, *Carbohydr. Polym.*, 2018, **186**, 184–191.
109. M. Martelli-Tosi, B. S. Esposto, N. Cristina da Silva, D. R. Tapia-Blácido and S. M. Jafari, Reinforced nanocomposites for food packaging, in *Handbook of Food Nanotechnology*, ed. S. M. Jafari, Academic Press, 2020, ch. 14, pp. 533–574.
110. S. Kumar, *et al.*, Biodegradable Hybrid Nanocomposite of Chitosan/Gelatin and Green Synthesized Zinc Oxide Nanoparticles for Food Packaging, *Foods*, 2020, **9**, 1143.
111. V. Ramji and M. Vishnuvarthanan, Influence of NiO Supported Silica Nanoparticles on Mechanical, Barrier, Optical and Antibacterial Properties of Polylactic Acid (PLA) Bio Nanocomposite Films for Food Packaging Applications, *Silicon*, 2020, **14**, 531–538.
112. I. Echeverría, M. E. López-Caballero, M. C. Gómez-Guillén, A. N. Mauri and M. P. Montero, Active nanocomposite films based on soy proteins-montmorillonite-clove essential oil for the preservation of refrigerated bluefin tuna (Thunnus thynnus) fillets, *Int. J. Food Microbiol.*, 2018, **266**, 142–149.
113. M. Mozaffarzogh, A. Misaghi, Y. Shahbazi and A. Kamkar, Evaluation of probiotic carboxymethyl cellulose-sodium caseinate films and their application in extending shelf life quality of fresh trout fillets, *LWT*, 2020, **126**, 109305.
114. D. De, M. S. Naga Sai, V. Aniya and B. Satyavathi, Strategic biorefinery platform for green valorization of agro-industrial residues: A sustainable approach towards biodegradable plastics, *J. Clean. Prod.*, 2021, **290**, 125184.
115. J. Xu, *et al.*, Amylose/cellulose nanofiber composites for all-natural, fully biodegradable and flexible bioplastics, *Carbohydr. Polym.*, 2021, **253**, 117277.
116. L. S. F. Leite, *et al.*, Eco-friendly gelatin films with rosin-grafted cellulose nanocrystals for antimicrobial packaging, *Int. J. Biol. Macromol.*, 2020, **165**, 2974–2983.
117. L. S. F. Leite, C. M. Ferreira, A. C. Corrêa, F. K. V. Moreira and L. H. C. Mattoso, Scaled-up production of gelatin-cellulose nanocrystal

bionanocomposite films by continuous casting, *Carbohydr. Polym.*, 2020, **238**, 116198.
118. Z. Yu, *et al.*, Effect and mechanism of cellulose nanofibrils on the active functions of biopolymer-based nanocomposite films, *Food Res. Int.*, 2017, **99**, 166–172.
119. Y. Han, M. Yu and L. Wang, Soy protein isolate nanocomposites reinforced with nanocellulose isolated from licorice residue: Water sensitivity and mechanical strength, *Ind. Crops Prod.*, 2018, **117**, 252–259.
120. Y. Han, M. Yu and L. Wang, Soy protein isolate nanocomposites reinforced with nanocellulose isolated from licorice residue: Water sensitivity and mechanical strength, *Ind. Crops Prod.*, 2018, **117**, 252–259.
121. G. Mondragon, C. Peña-Rodriguez, A. González, A. Eceiza and A. Arbelaiz, Bionanocomposites based on gelatin matrix and nanocellulose, *Eur. Polym. J.*, 2015, **62**, 1–9.
122. J. A. García-Ramón, *et al.*, Morphological, barrier, and mechanical properties of banana starch films reinforced with cellulose nanoparticles from plantain rachis, *Int. J. Biol. Macromol.*, 2021, **187**, 35–42.
123. R. A. Ilyas, *et al.*, Oxygen permeability properties of nanocellulose reinforced biopolymer nanocomposites, *Mater. Today Proc.*, 2022, **52**, 2414–2419.
124. A. Khezrian and Y. Shahbazi, Application of nanocompostie chitosan and carboxymethyl cellulose films containing natural preservative compounds in minced camel's meat, *Int. J. Biol. Macromol.*, 2018, **106**, 1146–1158.
125. F. Talebi, *et al.*, Incorporation of spice essential oils into poly-lactic acid film matrix with the aim of extending microbiological and sensorial shelf life of ground beef, *LWT*, 2018, **96**, 482–490.
126. M. A. Nouri Ala and Y. Shahbazi, The effects of novel bioactive carboxymethyl cellulose coatings on food-borne pathogenic bacteria and shelf life extension of fresh and sauced chicken breast fillets, *LWT*, 2019, **111**, 602–611.
127. O. Moreno, À. Gil, L. Atarés and A. Chiralt, Active starch-gelatin films for shelf-life extension of marinated salmon, *LWT*, 2017, **84**, 189–195.
128. J. Menezes and K. A. Athmaselvi, Report on Edible Films and Coatings, in *Food Packaging and Preservation*, ed. A. M. Grumezescu and A. M. Holban, Academic Press, 2018, ch. 5, pp. 177–212.
129. S. Benjakul, A. Singh, L. Chotphruethipong and A. Mittal, Protein-polyphenol conjugates: Preparation, functional properties, bioactivities and applications in foods and nutraceuticals, in *Advances in Food and Nutrition Research*, ed. D. Granato, Academic Press, 2021, ch. 8, vol. 98, pp. 281–320.
130. L. R. Amado, K. D. S. Silva and M. A. Mauro, Effects of interactions between soy protein isolate and pectin on properties of soy protein-based films, *J. Appl. Polym. Sci.*, 2020, **137**, 48732.

131. J. Cheng and L. Cui, Effects of high-intensity ultrasound on the structural, optical, mechanical and physicochemical properties of pea protein isolate-based edible film, *Ultrason. Sonochem.*, 2021, **80**, 105809.
132. S. Jiang, *et al.*, Modifying the physicochemical properties of pea protein by pH-shifting and ultrasound combined treatments, *Ultrason. Sonochem.*, 2017, **38**, 835–842.
133. X. Tian, *et al.*, Evaluation of a novel nano-size collagenous matrix film cross-linked with gallotannins catalyzed by laccase, *Food Chem.*, 2021, **351**, 129335.
134. N. Reddy, R. Reddy and Q. Jiang, Crosslinking biopolymers for biomedical applications, *Trends Biotechnol.*, 2015, **33**, 362–369.
135. T. Zhang, *et al.*, Modulating physicochemical properties of collagen films by cross-linking with glutaraldehyde at varied pH values, *Food Hydrocolloids*, 2022, **124**, 107270.
136. I. Brink, A. Šipailienė and D. Leskauskaitė, Antimicrobial properties of chitosan and whey protein films applied on fresh cut turkey pieces, *Int. J. Biol. Macromol.*, 2019, **130**, 810–817.
137. L. M. Reyes, M. Landgraf and P. J. A. Sobral, Gelatin-based films activated with red propolis ethanolic extract and essential oils, *Food Packag. Shelf Life*, 2021, **27**, 100607.
138. J. Lu, *et al.*, Optimization of heat-sealing properties for antimicrobial soybean protein isolate film incorporating diatomite/thymol complex and its application on blueberry packaging, *Food Packag. Shelf Life*, 2021, **29**, 100690.
139. F. V. R. Castro, M. A. Andrade, A. Sanches Silva, M. F. Vaz and F. Vilarinho, The Contribution of a Whey Protein Film Incorporated with Green Tea Extract to Minimize the Lipid Oxidation of Salmon (Salmo salar L.), *Foods*, 2019, **8**, 327.
140. C. Vilela, *et al.*, A concise guide to active agents for active food packaging, *Trends Food Sci. Technol.*, 2018, **80**, 212–222.
141. S. Yildirim, *et al.*, Active Packaging Applications for Food, *Compr. Rev. Food Sci. Food Saf.*, 2018, **17**, 165–199.
142. F. M. Yang, *et al.*, Effect of Nano-Packing on Preservation Quality of Fresh Strawberry (Fragaria ananassa Duch. cv Fengxiang) during Storage at 4 °C, *J. Food Sci.*, 2010, **75**, C236–C240.
143. S. T. Hammond, *et al.*, Food Spoilage, Storage, and Transport: Implications for a Sustainable Future, *BioScience*, 2015, **65**, 758–768.
144. D. Kowalczyk, *et al.*, Release kinetics and antimicrobial properties of the potassium sorbate-loaded edible films made from pullulan, gelatin and their blends, *Food Hydrocolloids*, 2020, **101**, 105539.
145. E. Abdollahzadeh, A. Nematollahi and H. Hosseini, Composition of antimicrobial edible films and methods for assessing their antimicrobial activity: A review, *Trends Food Sci. Technol.*, 2021, **110**, 291–303.
146. M. H. Lee, S. Y. Kim and H. J. Park, Effect of halloysite nanoclay on the physical, mechanical, and antioxidant properties of chitosan films incorporated with clove essential oil, *Food Hydrocolloids*, 2018, **84**, 58–67.

147. S. Ghani, H. Barzegar, M. Noshad and M. Hojjati, The preparation, characterization and in vitro application evaluation of soluble soybean polysaccharide films incorporated with cinnamon essential oil nanoemulsions, *Int. J. Biol. Macromol.*, 2018, **112**, 197–202.
148. A. Aydogdu, C. J. Radke, S. Bezci and E. Kirtil, Characterization of curcumin incorporated guar gum/orange oil antimicrobial emulsion films, *Int. J. Biol. Macromol.*, 2020, **148**, 110–120.
149. M. Pereda, A. G. Ponce, N. E. Marcovich, R. A. Ruseckaite and J. F. Martucci, Chitosan-gelatin composites and bi-layer films with potential antimicrobial activity, *Food Hydrocolloids*, 2011, **25**, 1372–1381.
150. A. Llorens, E. Lloret, P. A. Picouet, R. Trbojevich and A. Fernandez, Metallic-based micro and nanocomposites in food contact materials and active food packaging, *Trends Food Sci. Amp Technol.*, 2012, **24**, 19–29.
151. F. Tian, E. A. Decker and J. M. Goddard, Controlling lipid oxidation of food by active packaging technologies, *Food Funct.*, 2013, **4**, 669–680.
152. P. Jayasingh, D. P. Cornforth, C. P. Brennand, C. E. Carpenter and D. R. Whittier, Sensory Evaluation of Ground Beef Stored in High-oxygen Modified Atmosphere Packaging, *J. Food Sci.*, 2002, **67**, 3493–3496.
153. R. Domínguez, *et al.*, Active packaging films with natural antioxidants to be used in meat industry: A review, *Food Res. Int.*, 2018, **113**, 93–101.
154. A. Isenschmid, I. W. Marison and U. von Stockar, The influence of pressure and temperature of compressed CO2 on the survival of yeast cells, *J. Biotechnol.*, 1995, **39**, 229–237.
155. A. R. Mulakhudair, M. Al-Mashhadani, J. Hanotu and W. Zimmerman, Inactivation combined with cell lysis of Pseudomonas putida using a low pressure carbon dioxide microbubble technology, *J. Chem. Technol. Biotechnol.*, 2017, **92**, 1961–1969.
156. A. L. Holck, M. K. Pettersen, M. H. Moen and O. Sørheim, Prolonged shelf life and reduced drip loss of chicken filets by the use of carbon dioxide emitters and modified atmosphere packaging, *J. Food Prot.*, 2014, **77**, 1133–1141.
157. S. A. Cichello, Oxygen absorbers in food preservation: a review, *J. Food Sci. Technol.*, 2015, **52**, 1889–1895.
158. A. María Roberta, Oxygen scavenging films and coating of biopolymers for food application, in *Biopolymer Membranes and Films*, ed. M. A. de Moraes, C. F. da Silva and R. S. Vieira, Elsevier, 2020, ch. 22, pp. 535–551.
159. G. Singh, S. Singh, B. Kumar and K. K. Gaikwad, Active barrier chitosan films containing gallic acid based oxygen scavenger, *J. Food Meas. Charact.*, 2021, **15**, 585–593.
160. A. Cherpinski, *et al.*, Preparation and evaluation of oxygen scavenging nanocomposite films incorporating cellulose nanocrystals and Pd nanoparticles in poly(ethylene-co-vinyl alcohol), *Cellulose*, 2019, **26**, 7237–7251.
161. G. Sahoo, N. Sarkar and S. K. Swain, Antimicrobial Properties of Nanogold-Imprinted Starch Bionanocomposites, *Polym.-Plast. Technol. Eng.*, 2017, **56**, 334–345.

162. M. Kurtjak, M. Vukomanović and D. Suvorov, Antibacterial nanocomposite of functionalized nanogold and gallium-doped hydroxyapatite, *Mater. Lett.*, 2017, **193**, 126–129.
163. A. H. Virgili, *et al.*, Nanocomposite film with antimicrobial activity based on gold nanoparticles, chitosan and aminopropylsilane, *Surf. Coat. Technol.*, 2021, **415**, 127086.
164. Y.-C. Wang, L. Lu and S. Gunasekaran, Biopolymer/gold nanoparticles composite plasmonic thermal history indicator to monitor quality and safety of perishable bioproducts, *Biosens. Bioelectron.*, 2017, **92**, 109–116.
165. J. W. Rhim, L. F. Wang and S. I. Hong, Preparation and characterization of agar/silver nanoparticles composite films with antimicrobial activity, *Food Hydrocolloids*, 2013, **33**, 327–335.
166. Z. Wu, *et al.*, Multifunctional chitosan-based coating with liposomes containing laurel essential oils and nanosilver for pork preservation, *Food Chem.*, 2019, **295**, 16–25.
167. A. A. Oun and J.-W. Rhim, Preparation of multifunctional chitin nanowhiskers/ZnO-Ag NPs and their effect on the properties of carboxymethyl cellulose-based nanocomposite film, *Carbohydr. Polym.*, 2017, **169**, 467–479.
168. X. Guo, B. Chen, X. Wu, J. Li and Q. Sun, Utilization of cinnamaldehyde and zinc oxide nanoparticles in a carboxymethylcellulose-based composite coating to improve the postharvest quality of cherry tomatoes, *Int. J. Biol. Macromol.*, 2020, **160**, 175–182.
169. X. Zhang, *et al.*, Preparation of chitosan-TiO2 composite film with efficient antimicrobial activities under visible light for food packaging applications, *Carbohydr. Polym.*, 2017, **169**, 101–107.
170. M. A. Sani, A. Ehsani and M. Hashemi, Whey protein isolate/cellulose nanofibre/TiO_2 nanoparticle/rosemary essential oil nanocomposite film: Its effect on microbial and sensory quality of lamb meat and growth of common foodborne pathogenic bacteria during refrigeration, *Int. J. Food Microbiol.*, 2017, **251**, 8–14.
171. Y. Qiu, J. Fu, B. Sun and X. Ma, Sustainable nanocomposite films based on SiO_2 and biodegradable poly(3-hydroxybutyrate-co-3-hydroxyhexanoate) (PHBH) for food packaging, *E-Polym.*, 2021, **21**, 072–081.
172. D. Praseptiangga, N. Mufida, C. Panatarani and I. M. Joni, Enhanced multi functionality of semi-refined iota carrageenan as food packaging material by incorporating SiO_2 and ZnO nanoparticles, *Heliyon*, 2021, 7, e06963.
173. S. Tavakolian, H. Ahari, M. H. Givianrad and H. Hosseini, Improving the Barrier Properties of Food Packaging by Al_2O_3@TiO_2 & Al_2O_3@SiO_2 Nanoparticles, *Food Bioprocess Technol.*, 2021, **14**, 1287–1300.
174. S. M. M. Meira, G. Zehetmeyer, J. O. Werner and A. Brandelli, A novel active packaging material based on starch-halloysite nanocomposites incorporating antimicrobial peptides, *Food Hydrocolloids*, 2017, **63**, 561–570.

175. N. Devi and J. Dutta, Development and in vitro characterization of chitosan/starch/halloysite nanotubes ternary nanocomposite films, *Int. J. Biol. Macromol.*, 2019, **127**, 222–231.
176. S. Shankar, S. Kasapis and J.-W. Rhim, Alginate-based nanocomposite films reinforced with halloysite nanotubes functionalized by alkali treatment and zinc oxide nanoparticles, *Int. J. Biol. Macromol.*, 2018, **118**, 1824–1832.
177. B. Iamareerat, M. Singh, M. B. Sadiq and A. K. Anal, Reinforced cassava starch based edible film incorporated with essential oil and sodium bentonite nanoclay as food packaging material, *J. Food Sci. Technol.*, 2018, **55**, 1953–1959.
178. S. Cesur, C. Köroğlu and H. T. Yalçın, Antimicrobial and biodegradable food packaging applications of polycaprolactone/organo nanoclay/chitosan polymeric composite films, *J. Vinyl Addit. Technol.*, 2018, **24**, 376–387.
179. B. Shams, N. G. Ebrahimi and F. Khodaiyan, Development of Antibacterial Nanocomposite: Whey Protein-Gelatin-Nanoclay Films with Orange Peel Extract and Tripolyphosphate as Potential Food Packaging, *Adv. Polym. Technol.*, 2019, e1973184.
180. https://www.mordorintelligence.com/industry-reports, accessed 16th March 2022.

CHAPTER 6

Smart Food Sensing and IoT Technologies

B. T. W. PUTRA[a] AND B. KUSWANDI*[b]

[a] Laboratory of Precision Agriculture and Geo-informatics, Faculty of Agricultural Technology, University of Jember, Jl. Kalimantan 37 Jember, East Java, Indonesia; [b] Chemo and Biosensors Group, Faculty of Pharmacy, University of Jember, Jl. Kalimantan 37 Jember, East Java, Indonesia
*Email: b_kuswandi.farmasi@unej.ac.id

6.1 Introduction

Nowadays, the design, fabrication, and construction of smart sensors for food applications are one of the big challenges for food engineering researchers. They are crucial in the intelligent system, one of the cores of smart sensors technology and sensor systems. These concerns apply not only to the food sector but also to other sectors, where monitoring and diagnostics are needed, such as health, environment, *etc.* The ability to monitor continuously food integrity and control its environment inside food packages in real time can provide increased safety to the public, particularly concerning food safety and quality.[1,2] The capability to monitor food status and prevent packaging from leaching, including the ability to detect damage at an early stage, can reduce food loss and waste. Monitoring, controlling, and/or predicting the onset of food spoilage can provide advanced warning to allow for early food treatment or sorting. In addition, controlling and monitoring long-term food degradation, such as freshness or ripeness, is also crucial in the food sector.

Food Chemistry, Function and Analysis
Bio- and Nano-sensing Technologies for Food Processing and Packaging
Edited by Ashutosh Kumar Shukla

Published by the Royal Society of Chemistry, www.rsc.org

Naturally, food products over time will lose their properties and nutritional values rapidly, if they are not protected by packaging. Food packaging has a vital role in the food supply chain and its management system, as it can be used as a protective barrier from any contamination, and maintain food quality and safety.[3] However, traditional food packaging can only protect food products from the external environment and conditions, without being able to communicate with consumers regarding food conditions, such as freshness and safety.[4] Therefore, a novel food packaging system is required that allows to communicate with the consumer regarding its quality and safety. This could be achieved using a sensor, particularly employing an on-package sensor as smart/intelligent packaging.[5]

Smart packaging, especially intelligent packaging with on-package sensors, could support the efficiency and effectiveness of the food supply management system, *i.e.*, by enhancing food safety and food quality monitoring, including reducing food loss and waste. The intelligent packaging system, as it employs an on-package sensor, allows the food product to communicate with the user regarding the food condition, during transportation and storage, and to display *via* "monitoring" the food quality and safety.[6] Furthermore, this smart or intelligent packaging could be further developed as connected packaging, which allows consumers to interact with a food product *via* an on-package sensor or indicator and a package label or code that can be accessed using a mobile device. This connected packaging could be accessed remotely *via* IoT technology.

Rapid developments in sensor technology, wireless communication, microcontroller, microcomputer, IT, and IoT have the prospects for significant impacts on the food sector. To help in gathering the big data that is generated by a food monitoring system, a cloud system can be used to provide an adaptable, smart sensor, with rich features such as self-diagnosis and self-calibration capabilities. This chapter discusses smart food sensors that include smart packaging as an on-package sensor, since its function as a monitoring and analytical tool could be enhanced to be smarter with the help of the IoT *via* apps and artificial intelligence (AI). This chapter also summarizes previous research on sensor networks and provides a description of the smart sensors developed for food applications called connected packaging that allow users to interact with a food product remotely *via* an on-package sensor regarding food conditions, that can be activated with a mobile device, such as a smartphone, tablet, *etc.*

6.2 Smart Food Sensors

Generally, a sensor is a sensing device that is dedicated to obtaining information from a detectable phenomenon (from environment or object) and transforming it into a signal. Traditionally, the sensor can be divided into three parts: (i) a sensing element, (ii) a signal processor, and (iii) an interface for the read out.[7] While a smart sensor consists of sensors and a microcontroller or microprocessor, its intelligence capabilities can facilitate diagnostics, identification, or adaptation functions.[8] Today, most smart sensors are wireless, where radio frequency (RF) networks are used for data transmission. There are several network technologies for data communication, such as Wi-Fi and Bluetooth, that can be used to enable communication among devices using an intra-network and the internet. Commonly, these sensors use low radiated power to reduce costs related to the certified sensor. Thus, a smart sensor in this case has four vital components: (i) small size; (ii) processing unit, (iii) wireless, and (iv) low cost.[9] To put these things in their proper technical food perspective, the following describes food packaging that also has intelligent functions.

Food packaging is crucial in food production because packaging can protect a product from shocks, pressure, temperature, and other external elements. Smart packaging has been developed[10] based on biosensors and biopolymer nanocomposites. This concept is gaining attention in the industrial community due to food issues, *i.e.*, quality and safety. A biosensor integrated into food packaging enables real-time monitoring of food decomposition inside the packaging. Smart food packaging contributes to food quality, safety, and protection. This brings about added value through a longer shelf-life and affords multiple information, such as pH, temperature, humidity, and freshness. It even taps into the use of tracking devices. Examples of intelligent packaging are time–temperature indicator (TTI), temperature indicator, freshness indicator, ripeness indicator, barcode, QRcode, and radio frequency identification (RFID).

Intelligent packaging (IP) is described as a packaging system that facilitates decision-making to extend shelf-life, enhance food safety, improve quality, provide information, and also provide early warning of possible problems associated with food. This is due to it having intelligent functions such as sensing, detecting, recognizing, logging, tracing, communicating, and implementing, each of which is powered by scientific logic. Modified atmospheric packaging (MAP) is considered to be the basis for active packaging either using edible films or coatings used on food products, such as fruits and vegetables. MAP is useful in maintaining fruit quality for air storage. Active packaging includes various aspects such as physiological processes (*e.g.* the respiration of raw food), chemical processes (*e.g.* enzymic browning/oxidation reaction), physical processes (*e.g.* dehydration process), and microbiological aspects (*e.g.* food decomposition) that may affect the shelf-life.[11]

In a study conducted by Fuertes,[6] food packaging was developed by applying nanotechnology. Intelligent packaging plays an important role in food safety systems because it can monitor and provide information related to food, packaging, or environmental conditions influential to the packaging. Some examples of commercial intelligent packaging applications in the food industry include those listed in Table 6.1.

Table 6.1 Examples of commercial intelligent food packaging.[6]

Type of intelligent packaging	Brand	Company
Time and temperature indicators	Cook-Chex	Pymah Corp.
	Timestrip®	Timestrip Plc
	Colour-Therm	Colour-Therm
	MonitorMark™	3M™, Minnesota
	Onvu™	Ciba Specialty Chemical and FreshPoint
	Fresh-Check®	Temptime Corp.
	Thermax	Thermographic Measurements Ltd.
	CheckPoint®	Vitsab
Integrity indicators	Novas®	Insignia Technologies Ltd.
	Timestrip	Timestrip Ltd.
	Best-by™	FreshPoint Lab
	O_2 Sense™	FreshPoint Lab
	Ageless Eye®	Mitsubishi Gas Chemical Inc.
Freshness indicators	Fresh Tag	COX Technologies
	SensorQ®	DSM NV and Food Quality Sensor
	RipeSense	RipSense™ and ort Research
Radio frequency identification	Easy2log®	CAEN RFID Srl
	Intelligent Box	Mondi Pic
	CS8304	Convergence Systems Ltd.
	Temptrip	Temptrip LLC

6.2.1 Time and Temperature Indicators (TTI)

TTI is widely employed to monitor and transform the quality of food ingredients into certain indicators. In addition, TTI has been acknowledged as affordable, simple, and efficient. TTI plays an important role in maintaining food quality and safety, especially frozen food which requires a control point at the cooling temperature during the distribution process. Several types of TTI-based monitors are shown in Figure 6.1.

6.2.2 Integrity Indicators

Changes in gas composition during packaging can occur due to environmental factors, therefore a gas indicator is needed that can be used to control the composition of toxic gases resulting from packaged food decomposition, so as not to harm consumers.[12] For control measures, the indicator is evident through color changes in response to chemical or enzymatic reactions (Figure 6.2). The location of the indicator must be in direct contact with environmental gases inside the food packaging. In addition, for larger food packages, the gas sensor, especially the oxygen sensor, can be used for digital-based oxygen monitoring, and light-emitting diodes (LED) are used as a monitor.[13] The presence of oxygen in the packaging may

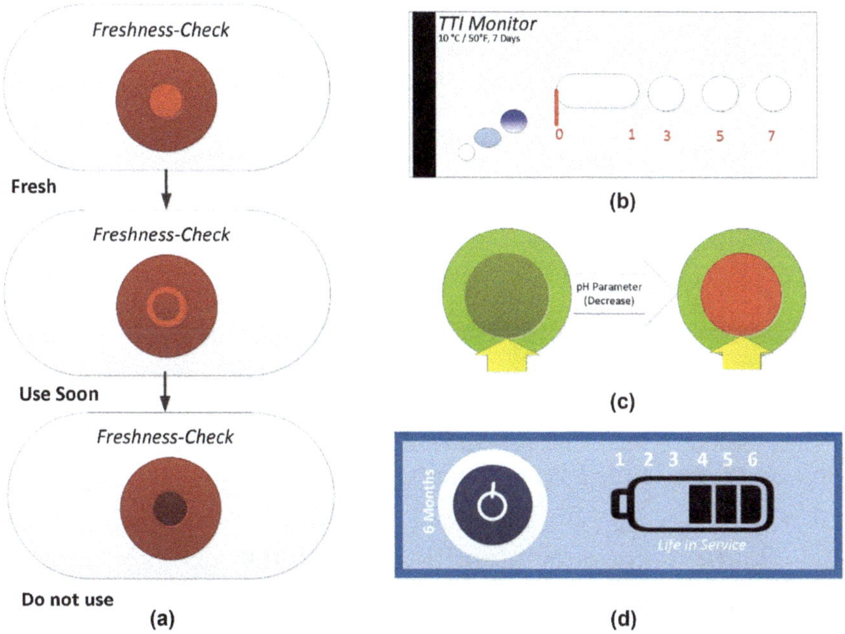

Figure 6.1 The TTI scheme: (a) freshness check; (b) monitor mark; (c) check point; (d) timestrip.

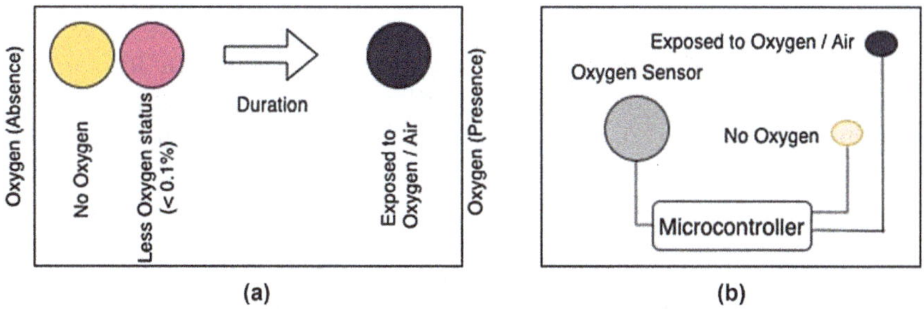

Figure 6.2 Indicator scheme for gas leakage: (a) paper-based monitor; (b) sensor-based monitor.

indicate that the food packaging is improperly sealed, leaked, or tampered with, either intentionally or unintentionally.

6.2.3 Freshness Indicator

The freshness indicator relates to a product quality which is generally found on the packaging label. This indicator detects changes concerning pH and gas composition, for example. The detector on this indicator will convert the signal into a color response because it can be directly related to the food freshness.[14] RipeSense™ is a smart sensor label that can indicate the

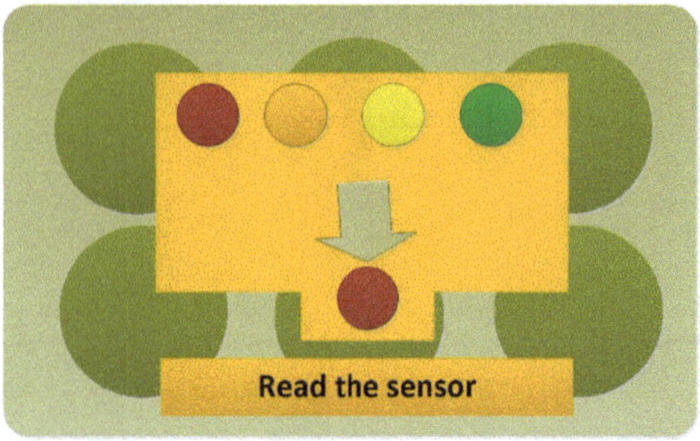

Figure 6.3 A sample of a ripeness sensor indicator on fruit.

ripeness level of fruit.[15] This sensor works based on a reaction with the fruit aroma according to the level of maturity (Figure 6.3).

6.2.4 Radio Frequency Identification (RFID)

RFID can identify information and place products with certain tags that utilize radio waves. RFID is classified into four types, including active, passive, semi-active, and semi-passive, depending on the power supply used for communication and other functions.[16] RFID can read information remotely. Active RFID can read details as far as approximately 90 meters away and has a battery that can communicate automatically. Passive tags have no internal power supply, so they cannot transfer any information until the RFID reader is activated. RFID systems are also classified based on the frequency used: low frequency (LF) at 125–134.2 kHz; high frequency (HF) at 13.56 MHz; ultra-high frequency (UHF) at 868–956 MHz; and active frequency or microwave frequency at 2.45 GHz.[17] The use of RFID in the food industry aims to control the freshness of meat, monitor *in situ* damage to fish, the freshness of fruits/vegetables, and real-time freshness systems of packaged dairy products.[18] The workflow of RFID is shown in Figure 6.4.

6.3 Internet of Things (IoT) in Food Monitoring

The stages in the food industry are interrelated with each other, ranging from upstream to downstream, or, in other words, from the seeds to the fork. These activities include on-farm research and development, food processing, packaging, distribution, advertising, and marketing. To support the food supply chain both upstream to downstream, advanced technologies play pivotal roles, such as the Internet of Things (IoT) supported by the use of artificial intelligence and big data.[19–21]

The Internet of Things (IoT) is one of the technological principles of Industry 4.0 integrated into an information technology (IT) infrastructure supporting data collection and distribution (Figure 6.5). Such integration is

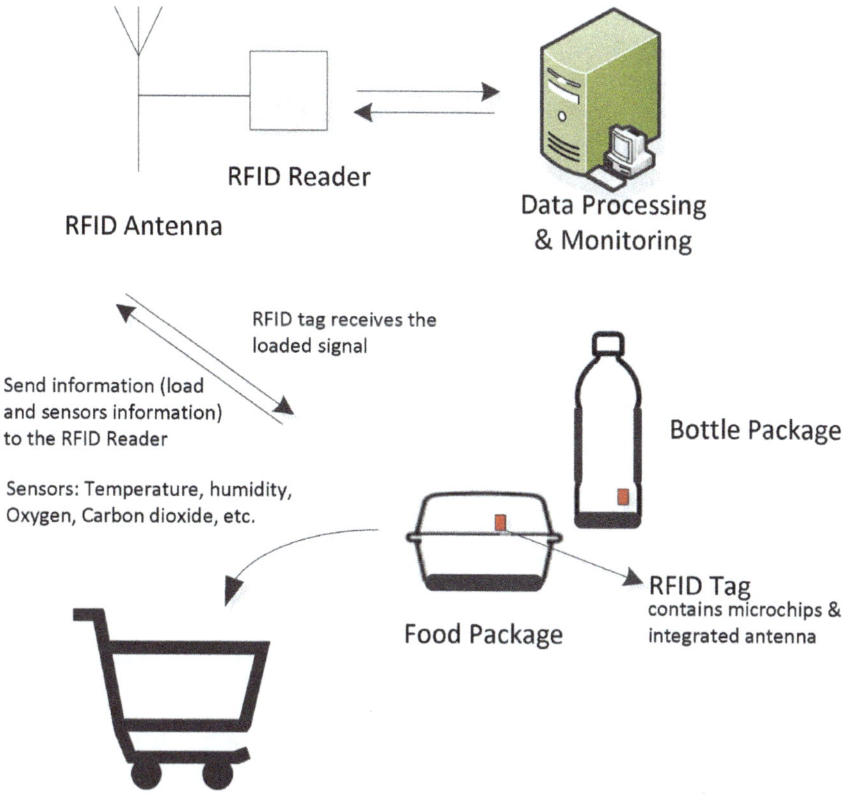

Figure 6.4 RFID for packaged fruit and drinks.

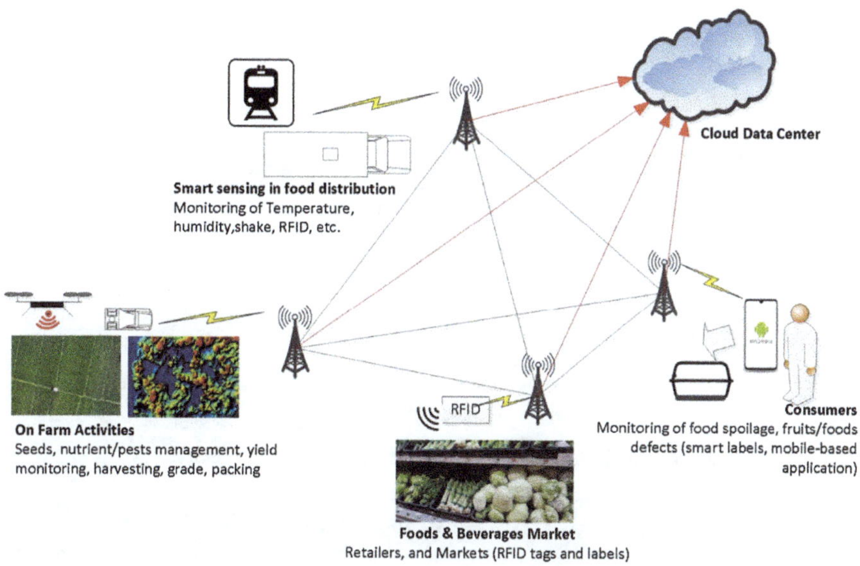

Figure 6.5 The Implementation of IoT in the Food Industry.

believed to significantly improve the efficiency and performance of manufacturing systems. The IoT supports the food industry through real-time data acquisition from multiple resources, including equipment, people, materials, and jobs. Real-time data retrieval is generally enabled by using sensors, RFID, and wireless communication standards. With this technology, the food monitoring process in the food industry can generate real-time and continuous information.[22]

In the food industry, the IoT can minimize the risk of food contamination. Different types of sensors are used to unify critical production status, delivery time, and especially temperature. The utilization of real-time temperature sensors enables the industry to closely monitor food safety and ensure effective chain management.[23]

In general, IoT resources can be categorized into three different layers.[24] The main architectural layers of the IoT encompass the sensing layer, network layer, and application layer, as illustrated in Figure 6.6. The vertical arrow shows the flow of technical implementation, starting from phenomenon detection to employing technology through specific applications. The horizontal arrow portrays steps involved in implementation and management.

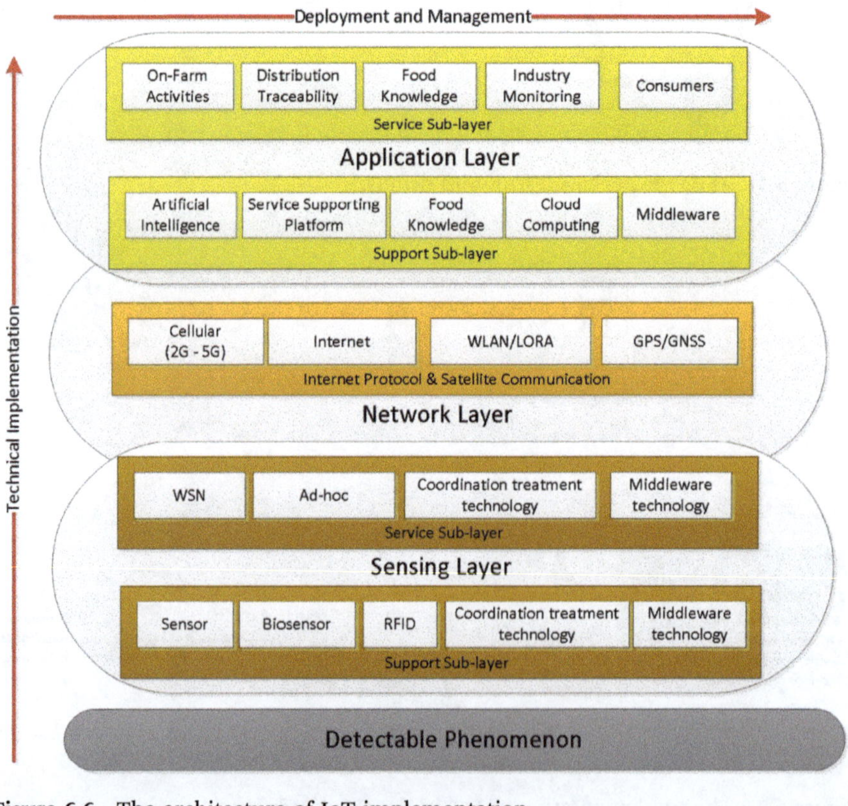

Figure 6.6 The architecture of IoT implementation.

These three layers can be described as follows:

a. *Sensing layer* relates to data taken from the environment from several stimuli, such as acoustic stimuli (*e.g.* wave, amplitude, phase, and spectrum), biological and chemical stimuli (*e.g.* gas and liquid concentration), electric stimuli (*e.g.* charge, current, permittivity, and conductivity), magnetic (*e.g.* flux and permeability), optical stimuli (*e.g.* reflection, color, scattering, and absorption), thermal stimuli (*e.g.* specific heat, thermal conductivity, and temperature), and mechanical stimuli (*e.g.* acceleration, strain, pressure, position, force, and velocity). To capture this information, sensors and biosensors are operationalized, where each sensor performs a specific function to sense a particular phenomenon in the environment. A number of sensors are at play, which includes RFID, camera, time, temperature, location, and machine. All data collected from multiple sensors at this layer are then transferred to the pre-processing stage.
b. *Network layer* aims to transmit information obtained from the sensing layer or data acquisition to be forwarded to the application layer through various network infrastructures, such as wire network, wireless sensor network (WSN), Bluetooth, Wi-Fi, and cellular phone connection (from 2G to 5G networks).
c. *Application layer* consists of several applications that provide the final information that can be displayed in real time through desktop, web, or mobile applications which report on specific information according to user roles (consumers, farmers, distributors, retailers, government, *etc.*). Some examples include modules of food quality and production efficiency. Once supplied with these data, these users are able to take the appropriate measures.

To complete the business requirements in agriculture, a single or several modules can be chosen by users. For example, they can use an on-farm application to find out processes from planting to harvesting. Another example is the quality traceability module, which is essential to trace agricultural products, qualities, and other information related to food (ingredients) to enhance user understanding.

6.3.1 IoT Sensors for Food Monitoring

The IoT system can carry out a monitoring process to ensure food quality. This will avoid contamination during storage and distribution. The analysis of temperature, humidity, and lighting that can affect food nutrition can be accessed by users through applications *via* smartphones. The web server is used as a data storage center connected to sensors that can be viewed and analyzed in real time. Users are notified about location information *via* messages if an emergency occurs during delivery. To retrieve food monitoring information, a sensing system is needed. To display data in the

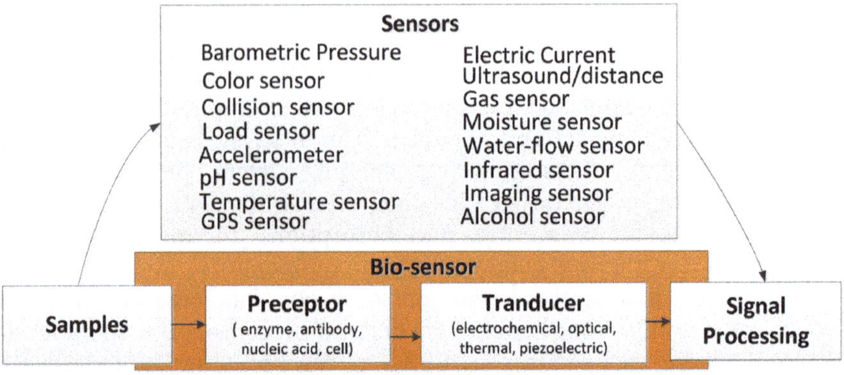

Figure 6.7 Several sensors for the Internet of Things.

monitoring process, a graphic user interface (GUI) is used *via* Bluetooth, Wi-Fi, or cellular phone.[25] The sensors used in the continuous food monitoring system include those presented in Figure 6.7.

There are several vital points to ponder in the deployment of sensors for food monitoring, namely the number of nodes, communication protocol, and the location of sensor node placement.[26] Concerning sensor node location, before implementing the sensor for monitoring, it is important to ascertain good environmental conditions, decent humidity levels, adequate communication signal coverage, and sensor quality. In this design, there are no special requirements regarding sensor placement that will be used for monitoring, so that it requires a state-of-the-art approach for reliable and sustainable information related to food monitoring.

In their implementation, sensors can be deployed using cable and wireless networks. The use of wireless networks is currently more flexible, but there are several obstacles to their implementation. The deployment of a sensor/sensor node may face difficulties in real conditions or in industrial environments due to several factors such as the actual location for placing the sensor/sensor node and its communication protocols. Poor environmental condition such as high humidity, high temperature, and inaccessible locations for signals make real deployments a very difficult task to plan.[27] While it is possible to find many approaches for floor planners and deployment of tools, most cases are extremely complicated to model and forecast before actual deployments. Figure 6.8 shows the architecture of sensor deployment and monitoring of real applications in food monitoring.

6.3.2 The IoT in Food Quality Monitoring

The application and integration of sensors with IoT technology allow continuous monitoring of food conditions and provide warnings when food property deviates from ideal parameters, therefore enabling users to accurately identify factors that damage food quality.[26] The system is developed to

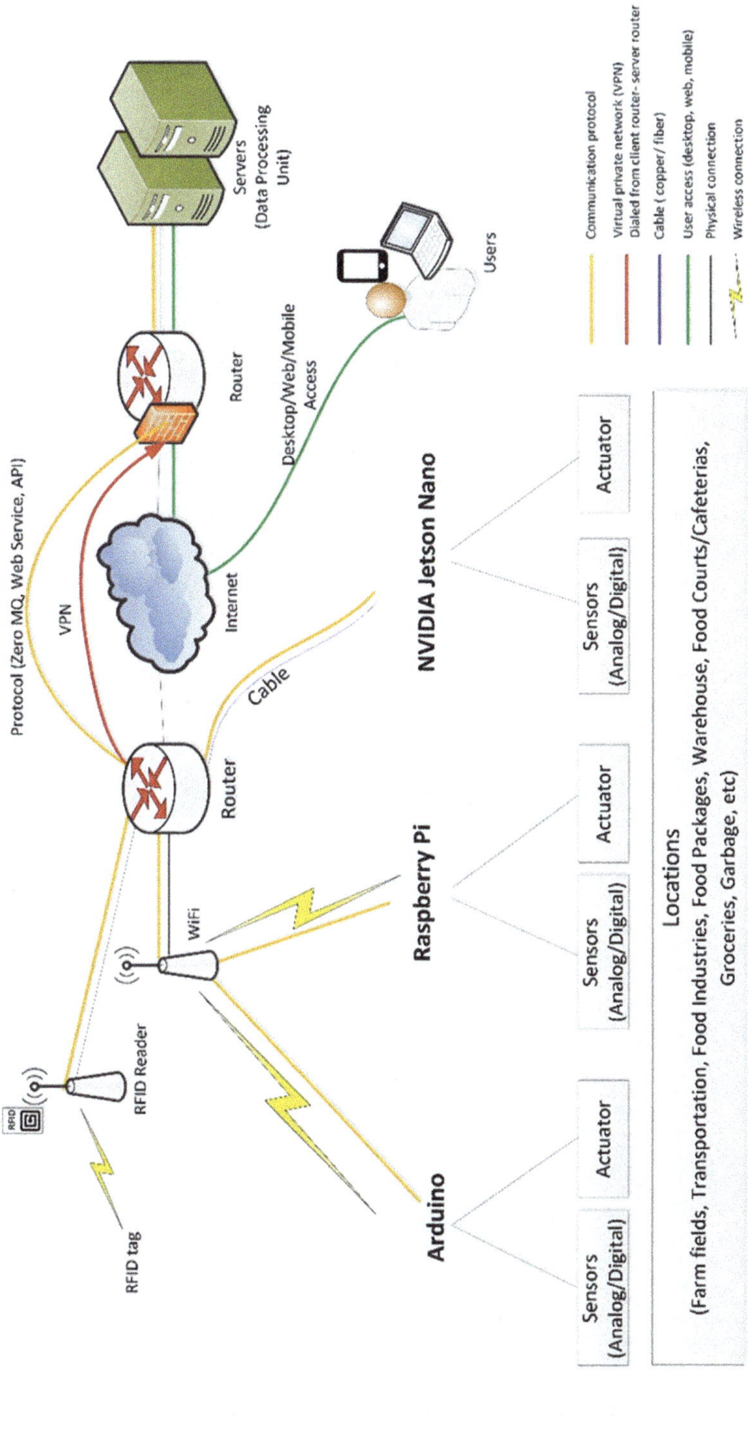

Figure 6.8 Sensors and network infrastructures.

detect spoilage and prevent contamination of food using multiple sensors. In evaluating and developing models to create a low-cost, effective, and efficient system, two methods are needed, namely destructive and non-destructive analysis. Destructive analysis is a method that utilizes chemical analysis through several processes such as destruction, heating, and digestion. It is used to determine the content of certain elements. However, this method takes a long time and is quite expensive. By contrast, the non-destructive method utilizes sensors to estimate certain parameters, phenomena, and elements. A sensor can only detect certain parameters or phenomena on a limited basis. As a result, it takes a variety of sensors that can perform multiple sensing functions in detecting certain phenomena. Many studies have developed various types of sensors.[17,27] Information obtained from sensors is analyzed and compared with data obtained from destructive analysis to create a model for system development.

The framework for food quality monitoring is illustrated in Figure 6.9. Samples of foods or beverages are analyzed by examining several parameters in the product, such as measuring color anomalies, and the amounts of cells, enzymes, and microorganisms. In addition, other sensors such as humidity sensors, pH sensors, temperature sensors, gas sensors, and chemical sensors are used to determine the extent of components damaging food quality.[28]

The research conducted by Tamilarasi *et al.*[29] involves the use of a wireless sensor network (WSN) to monitor food quality by combining biosensors and sensors. Another application was developed,[26] namely IoT Fog-Cloud,

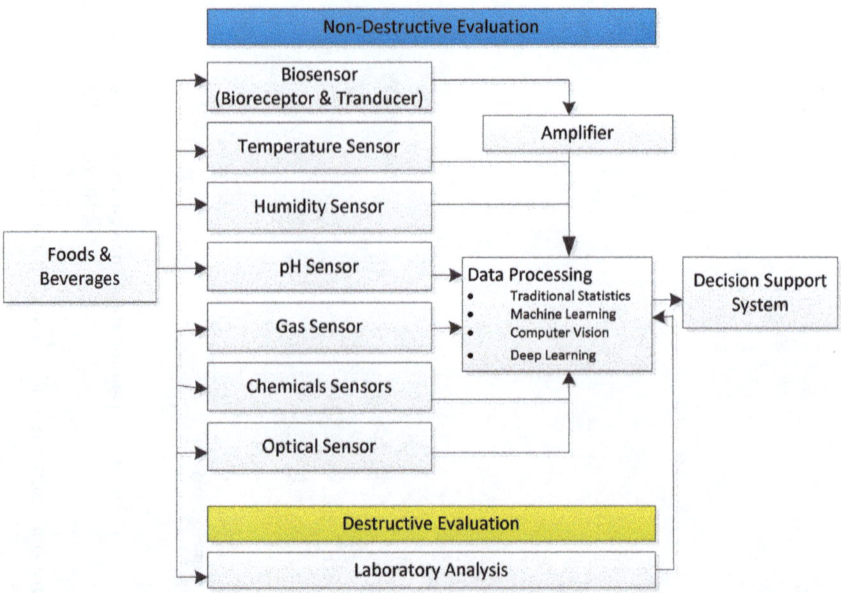

Figure 6.9 The system architecture for sensing the quality of foods and beverages.

which exemplifies how a sensor is used to evaluate food quality parameters. The IoT-Fog-Cloud integrated approach serves as an efficient framework for storing, processing, and analyzing information. The IoT components operative are wireless devices, preceptors, and RFID which can later transmit data over long distances. The technique presented consists of four phases, including acquisition, categorization, mining and extraction of data, as well as game-theoretic decision-modeling.

6.3.3 The IoT in the Food Supply Chain (FSC)

The long and complex FSC process causes food safety risks before the food is delivered to consumers, which potentially affects its safety and quality. The application of the IoT in FSC can significantly reduce the risk of counterfeit products, the use of counterfeit products, and food pathogens.[30] A number of sensors are used to track production sites, material sources, handling, and usage history. In addition, the integration of multiple sensors enables the monitoring of environmental conditions, such as temperature. One example of such integration, Hazard Analysis Critical Control Point (HACCP), allows digital and automatic data generation, which plays a role in the design of food safety solutions.

The application of the IoT in maintaining the quality and safety of food has been operationalized in several countries. The use of the IoT by combining data from the fresh vegetable supply chain to present an intuitive display for users, including end customers, thus helping users to decide when to buy food or monitor food supplies.[16]

The application of the IoT in FSC is designed to be integrated with machines and tools connected to an internet network. The IoT at FSC has an architecture consisting of four layers, namely the sensing layer, network layer, service layer, and application layer (Figure 6.10).

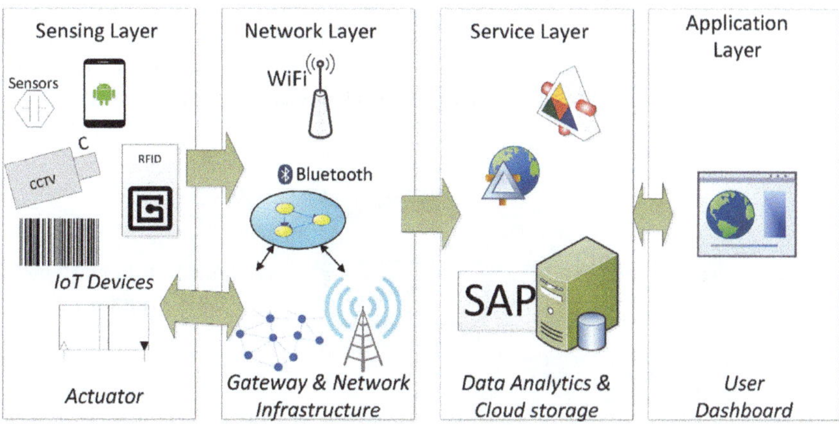

Figure 6.10 The architecture of the IoT in FSC.

a. *Sensing layer* relates to data taken from RFID sensors or cameras, such as time, environmental phenomena (*i.e.* temperature, moisture, humidity), location, and machine. All data taken at this layer are then transferred to the data pre-processing stage.
b. *Network layer* sends collected data at the sensing layer to the service layer through various networks, such as wireless sensor network (WSN), Bluetooth, Wi-Fi, LoRa.
c. *Service layer* functions to acquire data obtained from sensors, and perform storage, information analysis, and device management. This layer includes, for example, operational supports for data modeling, configuring and managing of IoT devices, and reporting information of business process management.
d. *Application layer* consists of several applications that generate the final information that can be displayed in real time, such as food quality modules and production efficiency. This layer plays a pivotal role in making the correct final decision.

6.3.4 The IoT in Maintaining Food Quality and Food Safety

Application of the IoT can maintain food quality and safety through various image processing technologies and the use of sensors. The sensors can continuously monitor the condition of food and provide a warning if there is a deviation in the product quality. The research conducted by Witjaksono *et al.*[31] is related to the development of mobile-based applications by utilizing optical sensors to check the freshness of shrimp and display the results on the consumer's cellphone. The food freshness level in real time (fresh, good, and spoiled) is displayed by reading the RGB color and comparing it with a paper pH reference.

In other applications, the gas sensor is one of the most important sensors applied to the food industry which aims to monitor food quality and safety (Figure 6.11). In a study conducted by Barandun *et al.*,[32] a paper-based electric gas sensor (PEGS) was developed to detect spoilage gases, such as ammonia and trimethylamine, in meat and fish products, so that they can inform consumers as to whether the product is consumable or not.

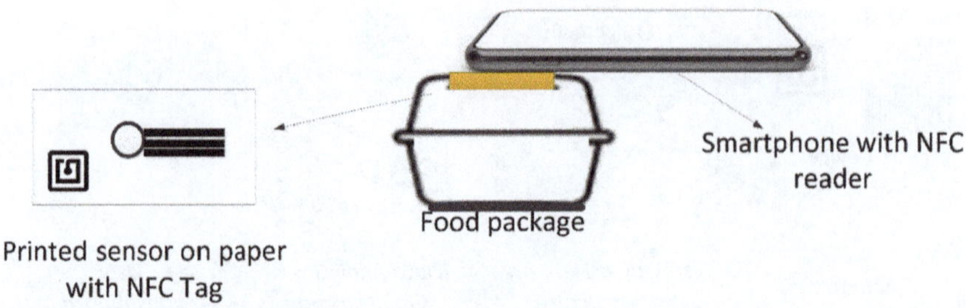

Figure 6.11 The detection of gas in food ingredients using a printed sensor.

6.3.5 The IoT in Food Waste

Lifestyles have changed significantly recently, with more people eating at cafes or malls. However, what is very unfortunate is that many food products are ordered and not consumed, although they may be still fit for consumption. The amount of leftover food produced can be quite large, and, if left unchecked, this can lead to various problems. In addition to food safety problems, it can cause environmental problems such as declining public health due to the increasing number of pathogenic microbes, blockage of sewers due to food waste directly discharged into sewers, groundwater and surface water pollution, soil pollution, and methane emissions from waste disposed of in landfills. To facilitate the management of food waste, an IoT-based waste management system can be a suitable option.[33] A comprehensive study (Figure 6.12) is needed for handling food waste that covers the entire process from waste collection and transportation, to waste treatment and final disposal.[34]

The development of this system could adopt the IoT to support supply chains, such as the use of smart trucks equipped with an integrated RFID system, security cameras, load sensors, GPS, and 2G/3G/4G/5G networks. These modules can be used to retrieve information related to truck latitude or longitude coordinates, routes, and operational status in real time. To transmit this set of information to the control center and to the vehicle's onboard controller, a GPRS/3G/4G/5G module or other cellular network is required.[22]

Surveillance cameras, for example, are placed in cafes or restaurants in conjunction with artificial intelligence to reveal the pattern of visitors in consuming food while estimating the amount of food waste and monitoring the process of collecting and transporting food waste in real time. In addition, the use of a weight sensor can monitor the amount of food waste that has been put into the trash bin, which subsequently informs the officer when the trash bin is almost full and ready to be collected. With a weight sensor,[34] food waste in the trash is weighed and the results of this sensor reading are sent to the server automatically in real time.

Another application in IoT-based waste treatment was investigated by Venkatesan *et al.*[33] The system developed was a landfill model designed on a prototype scale. Garbage was collected and stored for decomposition using citric acid which acts as a catalyst to accelerate the whole degradation process in a confined environment before delivery to a landfill. The smart package is equipped with sensors that will detect the presence of methane gas produced from waste degradation. As such, the amount of methane can be monitored in real time without directly accessing the landfill location. The methane produced is biogas, which potentially can be channeled and used as a renewable energy source. In addition, the by-products of these non-degradable materials can be used as organic fertilizers. The deployment of the sensor for monitoring food waste processing is shown in Figure 6.13. Data that have been obtained through the IoT system are analyzed to

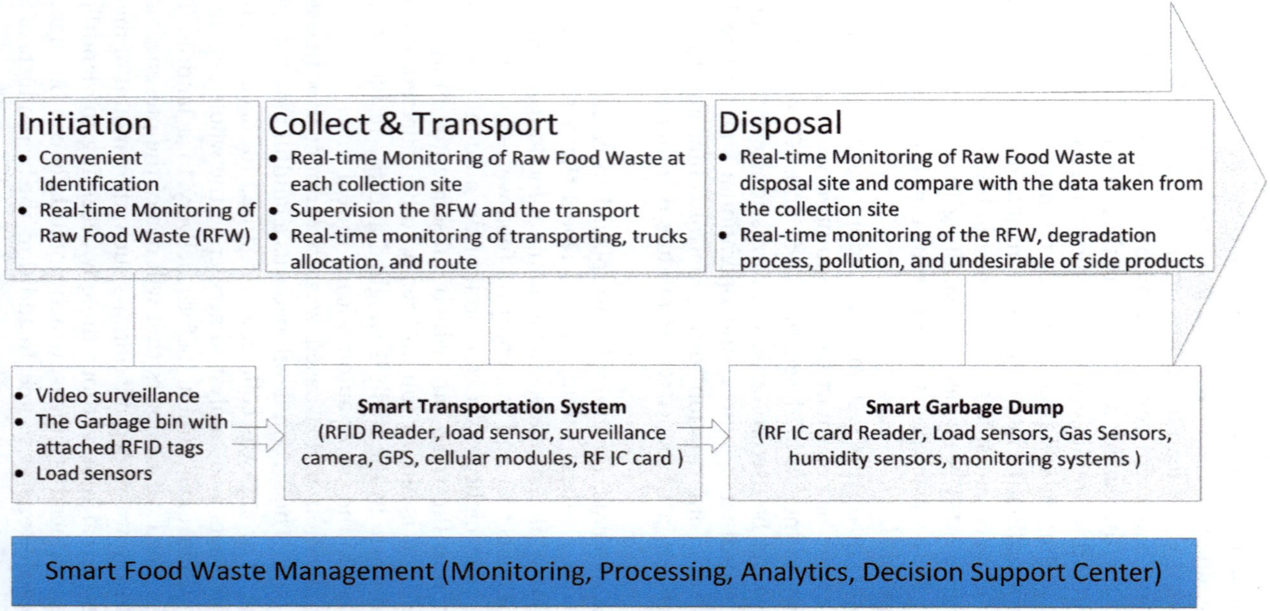

Figure 6.12 The overall architecture of an IoT-based system for waste processing.

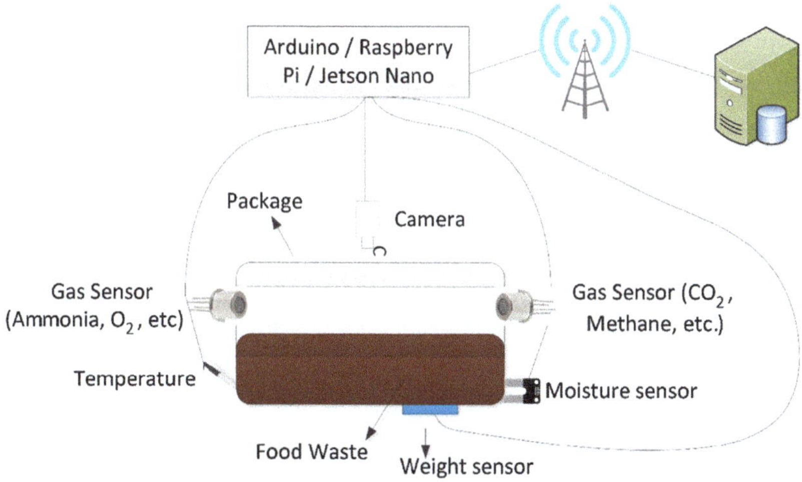

Figure 6.13 The architecture of good degradation monitoring in a laboratory.

produce reports for better food waste management operations and planning. This can help to bring greater focus into some of the intensive operational activities and increase the efficiency of using resources.

6.4 Computer Vision and Deep Learning in the Food Industry

Computer vision (CV) works as a sophisticated eye on a computer. Data processing in computer vision begins with data input, which includes some information. Afterward, the computer will process that information into a decision-making process and transform it into data output concerning pre-determined goals.[35] Computer vision belongs to the field of computer science that applies machines or devices to observe, understand, translate, or manipulate what has been observed.[36] The devices or machines commonly used to observe objects include cameras, Raspberry Pi modules, drones, and other devices for taking pictures and videos. While deep learning is a special subfield of machine learning, a new view of learning representations from data that emphasize layer learning may further enhance representation.[37] The relationship between artificial intelligence, machine learning, and deep learning is depicted in Figure 6.14.

Deep learning is a learning method with multiple levels of representation, obtained by constructing simple non-linear modules, each of which converts a representation at one level (starting with the raw input) into a representation at a higher and slightly more abstract level.[38,39] With the transformation composition, very complex functions can be put under analysis. For classification tasks, higher representation layers reinforce input aspects essential for discrimination and suppressing irrelevant variations.

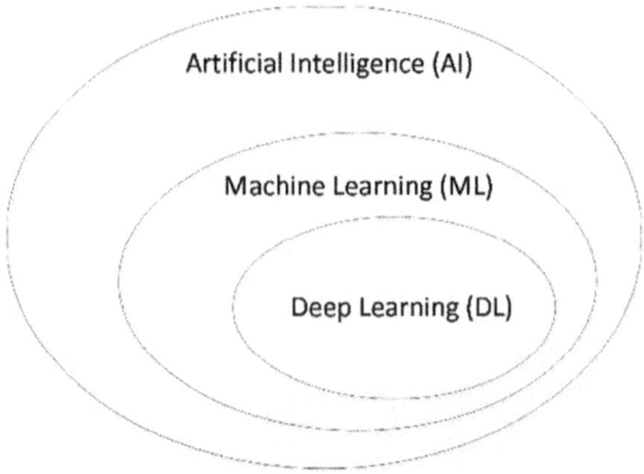

Figure 6.14 Artificial intelligence, machine learning, and deep learning.

The key aspect of deep learning is that these feature layers are not designed by human engineers, but are learned from the data using general-purpose learning procedures.[40] Deep learning is based on a specific type of learning mechanism. This is characterized by the efforts to create learning models at several levels, where the most in-depth level takes inputs and outputs from the previous level and transforms them into a more abstract format. Insights about this level of learning are adopted from how the brain processes information and responds to external stimuli.[41]

Computer vision implements deep learning techniques that can be utilized to simplify several tasks, such as image classification, object detection, object tracking, image manipulation, segmentation, and object recognition.[42] These simplifications allow a certain work to be completed more efficiently and accurately. Deep learning techniques serve as the brain or an important component that is closely related to computer vision.

Computer vision and deep learning have proven to be advanced technologies for big data processing in case problems, such as object discovery, speech recognition, object detection, and so on.[42] Even in the food and beverage industry, the use of computer vision and deep learning as data analysis tools to solve problems, including food and ingredient detection, calorie estimation, detection of the quality of fruit, vegetables, meat, and fishery products, food supply chains, and food contamination.[40] In this regard, careful attention needs to be drawn to specific problems, datasets, pre-processing methods, deep learning architectures and workflows used, outputs, and comparisons of several methods from each research conducted.

To detect an object in real time, a tool is needed to function as a picture or video recorder. One essential field of knowledge in this regard is computer vision. This works like the human eye by looking at an object with the aid of an image or video capture device (camera, webcam, drone, *etc.*), which subsequently leads to interpreting the information obtained using a computer.[35] Deep learning (DL) can help to detect, classify, and control the

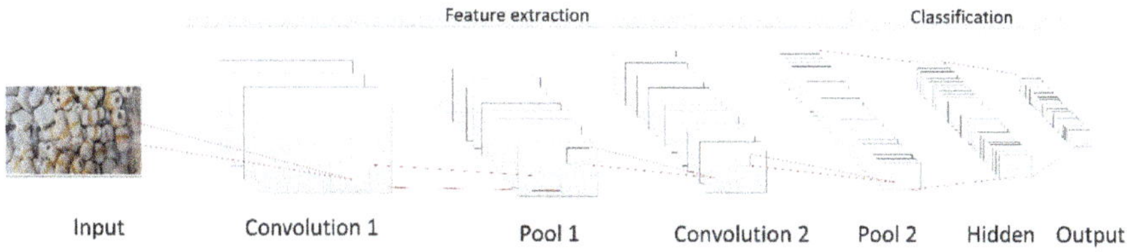

Figure 6.15 The architecture of a convolutional neural network (CNN).

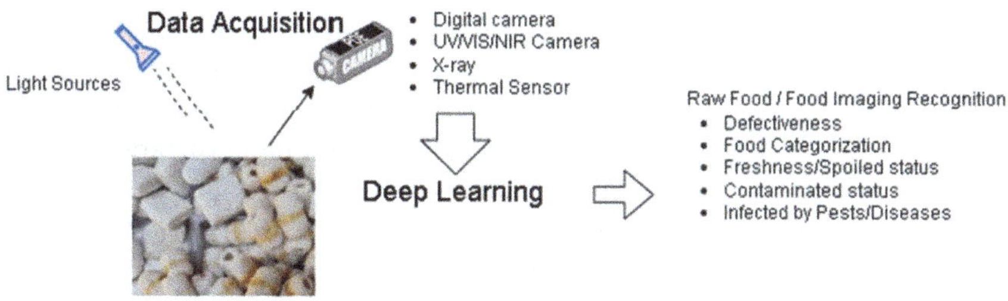

Figure 6.16 Raw food/food imaging recognition using deep learning.

quality of food products. The designer of DL architecture used for the detection or classification of an object is the convolutional neural network (CNN) (Figure 6.15). The results of DL modeling can be implemented in the food industry by utilizing the IoT. The following is an example of the application of CV and DL in the food industry that can be applied for automatically sorting dates based on the level of maturity. Based on research conducted by Nasiri *et al.*,[36] the maturity level of dates ranges from *Kimri*, *Khalal*, *Rutab*, to *Tamar*. This fruit is generally consumed when it is in the *Khalal*, *Rutab*, and *Tamar* stages.

The other DL applications can be used to control grain quality, freshness/ spoiled status, contaminated status, and pest/disease status (Figure 6.16). DL is used to detect the pest *Sitophilus zeamais* on infected and uninfected corn kernels with the help of X-ray images to control pest attacks.[38] The data used are trained and validated with a 70 : 30 ratio of all images, and the image size is resized to 160 × 160 pixels. The approach used requires CNNs to identify infected and uninfected classes of pests on wheat maize using three architectures, namely Inception-ResNet-v2, Xception, and MobileNetV2.

6.5 Conclusions

This chapter has provided a brief review of smart food sensor technology, revealing a number of its challenges and opportunities. The recent advances in smart food sensors depend on the micro- or nanomaterials, sensor technology, and wireless communication used that allow for monitoring the quality and safety of food directly or indirectly to help users in measuring food preferences, quality and safety, including shelf-life and best-date consumable.

The integration of smart sensors in food packaging has created innovative methods for the intelligent packaging system. This innovation has been able to increase the shelf-life and safety of food, while reducing food loss and waste. Nowadays, smart sensor innovations for foods have been guided by consumer preferences and needs globally, such as the IoT, big data, including AI, CV, *etc.*, including their smart mobile devices and apps. In addition, further advances in smart sensors have been enabled from the use of nanotechnology, which is sensing technology in the nano-size. Novel smarter sensors for food will be released shortly with the coupling of smart sensors and the IoT, so that smart packaging will provide consumers with greater convenience, in terms of food preferences, ingredient, quality, and safety. Furthermore, smart food sensors could be made smarter by their various functions, such as tracking, tracing, authentication, online dietary qualities, nutritional index, halalness, *etc.*, as the sensor can be connected to food packaging to communicate with the food product *via* a smartphone that contains important information about the foods preferred by customers.

Acknowledgements

The authors gratefully thank the LP2M, University of Jember, for supporting this work.

References

1. B. Kuswandi, Y. Wicaksono, A. Jayus, L. Y. H. Abdullah and M. Ahmad, *Sens. Instrum. Food Qual. Saf.*, 2011, **5**, 137–146.
2. K. B. Biji, C. N. Ravishankar, C. O. Mohan and T. K. Srinivasa Gopal, *J. Food Sci. Technol.*, 2015, **52**, 6125–6135.
3. B. Kuswandi, Y. Wicaksono, Jayus, A. Abdullah, L. Y. Heng and M. Ahmad, *Sens. Instrum. Food Qual. Saf.*, 2011, **5**, 137–146.
4. B. Kuswandi, Jayus, R. Oktaviana, A. Abdullah and L. Y. Heng, *Packag. Technol. Sci.*, 2014, **27**, 69–81.
5. B. Kuswandi, *Ref. Modul. Food Sci.*, 2018, DOI: 10.1016/B978-0-08-100596-5.21876-3.
6. G. Fuertes, I. Soto, R. Carrasco, M. Vargas, J. Sabattin and C. Lagos, *J. Sensors*, 2016, 4046061.
7. C. Adley, *Foods*, 2014, **3**, 491–510.
8. H. Xia, Y. Qiao, J. Jian and Y. Chang, *Sensors*, 2014, **14**, 20843–20865.
9. B. Kuswandi and M. Moradi, *Trends in Beverage Packaging*, Elsevier, 2019, pp. 279–302.
10. A. Sobhan, K. Muthukumarappan and L. Wei, *Food Packag. Shelf Life*, 2021, **30**, 100745.
11. S. Bodbodak and Z. Rafiee, *Eco-Friendly Technol. Postharvest Prod. Qual.*, 2016, 77–125.
12. T. Janjarasskul and P. Suppakul, *Crit. Rev. Food Sci. Nutr.*, 2018, **58**, 808–831.

13. A. K. Sundramoorthy, T. H. Vignesh Kumar and S. Gunasekaran, *Graphene-Based Nanosensors and Smart Food Packaging Systems for Food Safety and Quality Monitoring*, Elsevier Inc., 2017.
14. B. Kuswandi, *Freshness Sensors for Food Packaging*, Elsevier, 2018.
15. http://www.ripesense.com (available 12/10/2016).
16. G. Alfian, M. Syafrudin, U. Farooq, M. R. Ma'arif, M. A. Syaekhoni, N. L. Fitriyani, J. Lee and J. Rhee, *Food Control*, 2020, **110**, 107016.
17. S. Piramuthu and W. Zhou, *REID for Food Quality, Safety, and Security*, 2016, 243–266.
18. G. Alfian, J. Rhee, H. Ahn, J. Lee, U. Farooq, M. F. Ijaz and M. A. Syaekhoni, *J. Food Eng.*, 2017, **212**, 65–75.
19. S. Jagtap, L. Duong, H. Trollman, F. Bader, G. Garcia-Garcia, G. Skouteris, J. Li, P. Pathare, W. Martindale, M. Swainson and S. Rahimifard, *Food Technol. Disruptions*, 2021, 175–211.
20. B. T. W. Putra, K. S. Indracahyana and B. Al. Fanshuri, *Microprocess Microsyst*, 2022, **91**, 104550.
21. L. Da Xu, W. He and S. Li, *IEEE Trans. Ind. Informatics*, 2014, **10**, 2233–2243.
22. S. Jagtap, G. Garcia-Garcia and S. Rahimifard, *Comput. Ind.*, 2021, **127**, 103397.
23. H. Vaikousi, C. G. Biliaderis and K. P. Koutsoumanis, *Int. J. Food Microbiol.*, 2009, **133**, 272–278.
24. L. Zheng, H. Zhang, W. Han, X. Zhou, J. He, Z. Zhang, Y. Gu and J. Wang *et al.*, *Internet of Things – Global Technological and Societal Trends from Smart Environments and Spaces to Green ICT*, 2011.
25. I. M. Perez de Vargas-Sansalvador, M. M. Erenas, A. Martínez-Olmos, F. Mirza-Montoro, D. Diamond and L. F. Capitan-Vallvey, *Talanta*, 2020, **216**, 120985.
26. M. Bhatia and T. A. Ahanger, *Pervasive Mob. Comput.*, 2021, **72**, 101304.
27. E. Quesada, M. V. Maigler, A. Barbado, J. Valverde, J. Portilla and T. Riesgo, *Proceedings of the 2014 29th Conference Design of Circuits and Integrated Systems DCIS 2014*, DOI: 10.1109/DCIS.2014.7035577.
28. E. Tan, W. Ng, R. Shao, B. Pereles and K. Ong, *Sensors*, 2007, **7**, 1747–1756.
29. N. EA, D. Tamilarasi, S. Sasikala, R. R. Nair and K. S. Uma, *Microprocess. Microsyst.*, 2021, 103972.
30. G. Rateni, P. Dario, F. Cavallo, G. Rateni, P. Dario and F. Cavallo, *Sensors*, 2017, **17**, 1453.
31. G. Witjaksono, A. A. Saeed Rabih, N. B. Yahya and S. Alva, *IOP Conf. Ser. Mater. Sci. Eng.*, 2018, **343**, 12023.
32. G. Barandun, M. Soprani, S. Naficy, M. Grell, M. Kasimatis, K. L. Chiu, A. Ponzoni and F. Güder, *ACS Sens.*, 2019, **4**, 1662–1669.
33. G. Venkatesan, R. Mithuna and S. Gandhimathi, *Mater. Today Proc.*, 2020, **33**, 2729–2734.
34. Z. Wen, S. Hu, D. De Clercq, M. B. Beck, H. Zhang, H. Zhang, F. Fei and J. Liu, *Waste Manag.*, 2018, **73**, 26–38.

35. M. Ruiz-Altisent, L. Ruiz-Garcia, G. P. Moreda, R. Lu, N. Hernandez-Sanchez, E. C. Correa, B. Diezma, B. Nicolaï and J. García-Ramos, *Comput. Electron. Agric.*, 2010, **74**, 176–194.
36. A. Nasiri, A. Taheri-Garavand and Y. D. Zhang, *Postharvest Biol. Technol.*, 2019, **153**, 133–141.
37. F. Chollet, *Deep learning with Python*, Simon and Schuster, 2021.
38. C. B. da Silva, A. A. N. Silva, G. Barroso, P. T. Yamamoto, V. Arthur, C. F. M. Toledo and T. D. A. Mastrangelo, *Foods*, 2021, **10**, 879.
39. D. Su, H. Kong, Y. Qiao and S. Sukkarieh, *Comput. Electron. Agric.*, 2021, **190**, 106418.
40. Y. Lecun, Y. Bengio and G. Hinton, *Nature*, 2015, **521**, 436–444.
41. J. Shin, Y. K. Chang, B. Heung, T. Nguyen-Quang, G. W. Price and A. Al-Mallahi, *Comput. Electron. Agric.*, 2021, **183**, 106042.
42. L. Zhou, C. Zhang, F. Liu, Z. Qiu and Y. He, *Compr. Rev. Food Sci. Food Saf.*, 2019, **18**, 1793–1811.

CHAPTER 7

Nano-engineered Sensors for Food Processing

CHARLES OLUWASEUN ADETUNJI,*[a]
FRANK ABIMBOLA OGUNDOLIE,[b] MODUPE DORIS AJIBOYE,[c]
JOHN TSADO MATHEW,[d] ABEL INOBEME,[e]
WADZANI PALNAM DAUDA,[f] SHAKIRA GHAZANFAR,[g]
OLOTU TITILAYO,[h] OLUGBEMI TOPE OLANIYAN,[i]
OLUWATOSIN ADEMOLA IJABADENIYI,[j]
OLUOPE OLUFEMI AJAYI[k] AND JULIANA BUNMI ADETUNJI[l]

[a] Applied Microbiology, Biotechnology and Nanotechnology Laboratory, Department of Microbiology, Edo University Iyamho, PMB 04, Auchi, Edo State, Nigeria; [b] Department of Biotechnology, Baze University Abuja, Nigeria; [c] Plant Science and Biotechnology, Faculty of Science, Federal University Oye Ekiti, Nigeria; [d] Department of Chemistry, Ibrahim Badamasi Babangida University Lapai, Niger State 9111014, Nigeria; [e] Department of Chemistry, Edo University Iyamho, Nigeria; [f] Crop Science Unit, Department of Agronomy, Federal University Gashua, Gashua P.M.B 1005, Yobe State, Nigeria; [g] Laboratory for Reproductive Biology and Developmental Programming, Department of Physiology, Edo University Iyamho, Nigeria; [h] Department of Biotechnology and Food Science, Durban University of Technology, Durban 4001, South Africa; [i] Laboratory for Reproductive Biology and Developmental Programming, Department of Physiology, Rhema University, Aba, Nigeria; [j] Department of Crop Protection, Federal University Gasua, Nigeria; [k] Department of Biochemistry, Edo State University Uzairue, formerly Edo University, Iyamho; [l] Department of Biochemistry, Osun State University, Nigeria
*Emails: adetunjicharles@gmail.com; adetunji.charles@edouniversity.edu.ng

Food Chemistry, Function and Analysis
Bio- and Nano-sensing Technologies for Food Processing and Packaging
Edited by Ashutosh Kumar Shukla

Published by the Royal Society of Chemistry, www.rsc.org

7.1 Introduction

Nanotechnology approaches in food are an emerging technology, with nanotechnology and nanoscience becoming new frontiers in this century. Compared to their usage in medicine administration and pharmaceuticals, their uses in agriculture and food are fairly recent. The potential of nanotechnology has led to innovative methods of application in various food industries in the 21st century, as described by several scientists and engineers.[1] Nanoscale devices, which incorporate proteins, membranes, DNA, and other natural biomolecules, are frequently developed to mimic nanodevices found in nature. Nanotechnology, which is based on reactive particles smaller than 100 nanometres, is expected to play a major role in the application and development of new energy and electronics technologies in the 21st century. Quantum mechanics governs the properties of nanoparticles rather than the physical principles that govern larger particles.[2]

Toxicity, solubility, strength, colour, and chemical reactivity and are just a few of the physical and chemical qualities that can vary greatly. Hundreds of consumer items, such as sunlight-diffracting cosmetics, transparent sunscreens, penetration-increased moisturizers, lasting varnishes and paints, stain and odour-repellent coverings, *etc.* all use engineered nanoparticles. A nanometre (nm) is one billionth of a metre, and so it is tiny even when compared to a human hair, which is 80 000 nm thick. The food market necessitates innovative technologies that permit fresh, tasty, and authentic food, and that improve freshness and extend the shelf life in an effort to maintain better market dominance in the various food processing industries.[3]

Nano-engineered sensors are employed in agriculture and food processing, storage of food products, nano-additives, cleaning, distribution, for pollutant detection, as well as the development of unique goods, which are all expected to benefit from these new products, along with their respective applications in food processing. In recent times, given the popularity gained by the application of nanotechnology in processing, food businesses are now open to various interesting possibilities.[4,5]

The field of nanotechnology combines chemistry, physics, biotechnology, and technology.[6] Several nanoscale components exist naturally in food products, these include the nanosized particles formed during digestion, fat globule micelles, and whey proteins. The scientific community has developed new technologies aimed at improving the shelf life of products, with several tools and methods designed to employ swift *in situ* analysis which has provided food packages that are environmentally safe and degradable due to their novel properties, such as mechanical properties, bioavailability, diffusivity features, among others.

Nanotechnology is applied in the area of food processing. Nanomaterials are employed as nutrient transporters, additives, antimicrobials, transducers in optoelectronics and as sensors for the evaluation of food quality.[7]

Nanostructured innovations employed in food processing have created promise for the improvements of food texture, consistency, nanostructured innovations employed in food processing have and bitter taste.[8] Because safe monitoring of food production and supply is essential, traditional laboratory equipment can only be used to evaluate a small number of samples, whereas nanomaterials have enabled the creation of a number of sensing devices for monitoring food at the source. Chemical transduction methods are developed mainly for sensing volatile substances; biomolecular recognition uses biosensors for greater selectivity in these systems. These biosensors have biological receptors for antibodies, aptamers, cells, enzymes, and so on that are used to detect a specific target, and its binding process can be carried out using electrochemical, optical, thermal, or mass processes.

The majority of biosensors discovered so far use electrochemical and optical processes, such as colour and electrical current movement.[9] There are biosensors that detect pathogens such as *E. coli* and *Salmonella*,[10] contaminants such as heavy metals,[11] residual animal care antibiotics,[12] residual pesticide residues, illegal additives,[13] mycotoxins, bisphenol, allergens, and antioxidants and others.[14,15] Canola oil, which is a nanosized micellar, provides delivery models for an array of substances including minerals, phytochemicals, vitamins, and a wide range of nanoceutical products as the delivery vehicles for vitamins and minerals. Varieties of nanoceuticals act as targeted delivery vehicles, such as a chocolate drink that is sweetened without the use of sugar, nano-based multivitamins, and a growing amount of nanosilver and nanogold supplements, and so on. This review summarizes recent studies on the nano-engineered sensors for food processing, focusing on a general overview, and their relevance and applications, and reports on the application of nano-engineered sensors in food processing.

7.2 Application of Nano-engineered Sensors in the Food Industry

Generally, nanotechnology is considered to be an innovative technology that allows involvement, development, and long-term effects in the medicine, agriculture, and food industries. Nanomaterials have the capacity to drive the quantitative and qualitative development of safe, good-quality perishable, healthier, and semi-perishable functional foods. Nanotechnologies outperform traditional food processing technologies in terms of reducing contamination, extending life span and, producing higher quality foods. Nanotechnology applications improve food bioavailability, consistency, taste, and texture, changing the particle size, suture formation, the surface charge of food nanomaterials, and cluster formation.[16] The current global population is about 6 billion people; as a result of this growing population, more efficient, safe, and sustainable food production is required, as well as food that is free of the harmful impacts of pathogenic organisms.

Sensor packaging, preservatives, and customized foods are applications of nanotechnology. The notion of customized food was based on the concept of as-required food, allowing customers to customize food based on their nutritional needs or preferences. In the food sector, nanotechnology has a number of beneficial effects that are predicted to expand in the near future. Due to their distinct and new features, nanomaterial applications will continue to have an economic impact on the food sector.

Nanomaterials are now being advanced in a way that will not just affect food flavour and food safety, and improve health. From food production to packing, processing, transportation, shelf life, and nutritional availability, this fast-evolving technology is now very important in all sectors of the food system. As a result, nanotechnology could help the food business to grow.[6] When nanoparticles within nanotechnology strategies are employed in the creation, processing, and packaging of food, it is referred to as nanofood. It should be stressed that this does not imply that the food has been atomically transformed or that it has been manufactured using nanomachines.

7.3 Nanotechnology in Packaging

Nanotechnology's industrial applications quickly improved meat packaging materials, ensuring product safety, nutritional and organoleptic attributes, as well as continuous monitoring throughout the shelf-life of products. Technology advances resulting in the development of nanosensors for smart and active packaging of products have tremendously aided the maintenance and improvement of food quality and the rise in their demand as a result of their safety. Gedarawatte *et al.*[17] in their research discovered that a bacterial nanocellulose which is loaded with nisin has antimicrobial properties and is effective against some bacteria associated with meat spoilage, they further concluded that this technology could be used to effectively package food materials.

Using okra mucilage with zinc oxide nanoparticles coated with carboxymethyl cellulose-based film to enhance the lifespan of chicken meat was successful. Pathogenic bacteria were discovered to be resistant to edible films which were created when turkey meat was preserved by coating with silver nanoparticles.[18] Furthermore, bacterial deterioration of packed chicken meat was minimized by polylactic acid nanocomposite films obtained from a mixture of cinnamon essential oil and nanostructures of silver–copper when compared to fresh meat.[19] Cushen *et al.*[20] reported the use of ICP-MS techniques for evaluation of the movement of AgNPs from the polyvinyl chloride (PVC) matrix into poultry meat. The acquired results revealed that the amount of AgNPs detected did not exceed the European Union's regulatory limits.

According to Khalaf *et al.*,[21] AgNPs containing edible films from pullulan also showed a good antibacterial effect and stability on the quality of turkey deli meats. Others, such as Fernández *et al.*,[22] found that AgNP absorption pads had an efficient microbiological preservation effect against *Staphylococcus aureus* and *Escherichia coli* poultry flesh. Impact assessment of poultry and pork products by Martinez-Abad *et al.*[23] showed the strong anti-*Listeria monocytogenes* potential of AgNPs containing ethylene-vinyl alcohol films. Additionally, the antibacterial capacity of polyethene with low density that is mixed with Ag and zinc oxide-based nanoparticles on chicken breast flesh was demonstrated against *Pseudomonas aeruginosa*, *Listeria monocytogenes* and *E. coli*.[24]

In nano-delivery systems, bioactives such as iron, curcumin, coenzyme Q10 (CoQ10), calcium, vitamins and so on have been extensively investigated. Several nano-delivery vehicles such as lipid-based nano-encapsulators, nanoparticles, biopolymeric nanoemulsions, nanolaminates, nanoemulsions, nanocarriers, nanofibres and association colloids, among others, are now in use. These nano-delivery technologies can boost bioactives' bioavailability through a variety of mechanisms. Through specialized delivery methods, nanoencapsulation can improve the bioavailability of bioactive substances following oral administration. This type of nanoencapsulation allows flavour release to be controlled at a certain time, while also preventing flavour degradation throughout storage and processing.[25]

Another interesting solution to food safety concerns can be addressed by nanosensors, which are now being engaged in detecting pesticide traces in a variety of foods. These are critical because such instruments are capable of detecting extremely small amounts of organic molecules, low concentrations of pathogens and other hazardous chemicals. These technologies are incredibly sensitive, quick to respond and capable of preventing harm.[26] Nanosensors are most commonly employed to identify pesticides such as organophosphate in fruits, aquatics and plants.

7.4 Application of Sensors in the Food Industry

The increased interest in the quality of food, along with its safety, necessitates the advancement of reliable and sensitive analytical methodologies, and also food quality and freshness preservation technology. The current state of biological and chemical sensors for smart packaging and food monitoring is reviewed here. Alongside examples of sensing designs, applications and related analytical properties for evaluating allergens, freshness markers, adulterants, toxicants as well as pathogens are explored. The use of packaging technologies might make it easier to minimize food waste, check food quality, improve overall food safety and increase shelf life.[27–37] Several sensors, on the other hand, are currently at the stage of development, and therefore require a lot of work before they can be used in real-world applications. Selectivity, sensitivity, safety and robustness of nanosensors associated with possible migration of contact in food must all be determined.[38]

The food processing sector faces a number of issues, one of which is the monitoring of nutritional quality and food safety. Traditional safety and quality analysis approaches are arduous, needing highly skilled personnel as well as being time-consuming. Because inappropriate storage and handling can lead to food poisoning, this approach cannot be relied upon. As a result, sensitive, dependable approaches enabling quick assessment of food safety and quality are required. In the food processing industry, this could also be avoided by adopting the sensor automating approach. Choosing the correct sensor systems for such a job is crucial to getting trustworthy findings. Sensors have to be flexible and durable, and have linearity of response, sensitivity, accuracy, repeatability, quick recovery and reaction times, reliability and stability, irrespective of the location, to meet these objectives. The capacities, restrictions and applicability of the device for the target purpose must all be carefully considered when selecting sensors. The shelf-life study, process control, authenticity assessment, freshness evaluation and other quality management investigations are all common uses of sensors in the production process. In the food industries, several types of sensors are utilized due to their high versatility and level of application or functionality. Some include temperature-monitoring sensors (infrared sensor, resistance temperature detector, thermocouple as well as a thermistor), location-based sensors (capacitive, ultrasonic as well as inductive), humidity sensors (gravimetric, capacitive, piezoresistive, magnetoelastic and resistive sensors), and biosensors (conductometer, amperometric, potentiometric biosensor and thermometric biosensor), pressure sensors, chemosensors, torque sensors, E-tongue taste sensors and freshness sensors. Sensors play a significant role in the identification or detection of pathogens all through the food production processes, and also in food processing industries, and help in improving the safety, quality, profitability and production of food.[39]

In the food sector, biological and environmental management is critical. Unwanted microbial deterioration in contemporary food processing is dangerous to customers' well-being, and costs processors and merchants a lot of money by leading to food waste, as well as being a concern for the global supply of food. The major aim of quality control in food processing is the identification and rectification of defects in the final products. This is achieved through the identification and elimination of various sources of quality challenges. It would also enable the food manufacturer to halt production, minimizing the damage to only a portion of the current sequence, resulting in significant cost savings in avoiding the use of raw materials, based on product valuation, packaging, as well as significantly reducing waste. Sensor arrays are more affordable and versatile than existing systems, however, despite their advantages, there is a need to make the current system an advanced process control tool through its proper integration into the processing systems. This would help in actualization of the various analytical targets such as high sensitivity, portability, high speed of utilization and low cost.[40]

One of the most pressing difficulties facing the food industry is the need for cost-effective and rapid technologies to identify pathogens and allergenic components in food. Easy and swift detection of pesticide residues in food, microbe and allergy detection, contaminant detection, verification of the content of a product, monitoring the conversion of raw materials and product freshness monitoring are among the possible uses of biosensors. The food and agricultural sectors constantly carry out quality assurance tests and monitor chemical and microbiological parameters to maintain their high standards and ensure food safety through qualified personnel. These analyses can be time-consuming and/or costly, requiring steps such as sample pretreatment and sometimes extraction steps. Most of the drawbacks mentioned earlier can easily be circumvented through a less time-consuming method which is standardized, non-destructive, inexpensive and also fast as results can be obtained in a timely fashion using a biosensor. Biotechnologies now have the ability to create an imaginative renaissance in the farming and food areas, thereby assisting with tackling these issues.[41]

Food industries today have constantly increasing demands for effective quality management solutions to meet regulatory and customer standards, as well as to increase quality sorting, production feasibility, time and cost savings and automation. Expensive and time-consuming analyses are carried out regularly as quality control by food and agricultural industries, although these could easily be addressed with the use of biosensors which will save time and avoid prolonged test periods, because they are non-destructive, cheap, quick, and provide effective and reliable results. Given all of these characteristics of the food and dairy industries, the invention and application of biosensors have proven to be a blessing. The basic ideas and utilization of biosensors in industries such as the dairy, brewery, wine and food industries have been highlighted in this study. Biosensors already have the potential to bring about an innovative breakthrough in processing industries, especially the food, wine, dairy and other manufacturing industries.[42]

Food freshness is one of the most significant features that customers should evaluate whenever purchasing food. Several consumers, on the other hand, have difficulty determining the amount of freshness without the need for gadgets or equipment. As a result, the creation of a food freshness sensor has emerged as among the potential analysis techniques, offering greater advantages over other instruments such as high-performance liquid chromatography (HPLC) and gas chromatography–mass spectrometry (GC–MS). Food freshness sensors allow for speedier examination, while also being non-destructive to the food sample. The creation and functioning of many different sensors enabling food freshness detection, including optical sensors, electrochemical sensors, pH sensors along with potentiometric sensors, are the subjects of this research. There are various primary components that will be stressed for every sensing device, including the chemicals employed, type of samples, sensor performance and immobilization matrix.

Biosensors are not often employed for food microbiological analysis, although they have a great deal of potential for detecting microbial infections, including those related to poisons. For both field and laboratory analyses, they provide mobility, quick or real-time detection, and also multi-pathogen identification. For microbial investigation of foodborne pathogens including *S. aureus*, *C. perfringens*, *Campylobacter*, *E. coli*, norovirus, *L. monocytogenes*, and several species of *Salmonella*, along with microbial toxins like mycotoxins and staphylococcal enterotoxins, various applications are available. These technologies are based on a biological control agent in conjunction with such a process control device. Increasing the output or enhancing the conservation of farm commodities is a typical goal for farmers. However, the usage of xenobiotics plainly poses a risk to buyers. Furthermore, the food business, control laboratories and regulatory authorities are all concerned about safety concerns. Also ,with complementary approaches, instrumental analytical processes to identify quality characteristics and safety concerns are needed, such as immunoassays, bioassays, lateral flow devices and chemical assays, that could be used in delocalized research or as quick screening techniques.[43]

7.5 Authors that Have Worked on the Utilization of the Nanoengineered Sensor in Food Industry

Nanotechnology is a critical technology whose advancement has resulted in the development, contribution and long-term effects in the agriculture and food industries and recently, the medical/biomedical industries. Nanomaterials have the potential to drive the development of functional foods that are high quality or semi-perishable, healthier and safe. Nanotechnologies outperform traditional food processing technologies in terms of reducing contamination, increasing life span and producing higher quality food. Nanotechnology applications improve food bioavailability, consistency, taste and texture, change particle size, suture formation, and the surface charge of food nanomaterials and cluster formation.[16]

Generally, some of the applications of nanotechnology include foods on-demand additives, and smart packaging of foods that are on-demand and allows customers to customize their foods based on their nutritional needs or preferences. In the food sector, nanotechnology has numerous applications and this is anticipated to expand in the near future due to the increasing number of studies that are being carried out in this regard. Due to their distinct and new features, nanomaterial applications will continue to have an economic impact on the food sector. New nanomaterials are being created not only to influence food flavours but also to address food safety and provide several well-being benefits. From food production to packing, processing, transportation, shelf life and nutritional availability, this rapidly advancing innovation is concerned with each feature of the food framework. As a result, nanotechnology could help the food business grow.[6]

7.6 Nanotechnology in Chitosan

The demand for wholesome and safe food that uses few synthetic ingredients is steadily rising. Due to the fact that synthetic plastic polymers are harmful to the environment, considerable work has gone into evaluating different sources and types of bio-based natural polymers as potential replacement for synthetic plastic packaging. The use of chitosan, a natural polymer obtained through the deacetylated derivative of chitin obtained from the scales of fish and shrimps, and the exoskeleton of crustaceans is fast becoming an alternative due to its properties making it an appealing biopolymer for use in packaging foods and also as an additive for preservation. It has been reported by Zhu *et al.*[44] to be effective with a combination of cross-linked bilayers and oxidized form of ferulic acid to create films that form a bilayer which is made up of alginate and chitosan. An in-depth analysis was carried out on selected properties such as the optical and permeability capacity and the mechanical components of the different films created. The most suitable concentration was further obtained for subsequent studies. It was further observed that while there is no significant difference in their thermal stability, there was a more uniform and smoother appearance of the cross-linked bilayer film compared to the uncross-linked bilayer film, according to electron microscopy and thermogravimetric analyses.

The use of active bilayer films for the preservation of postharvest cherry tomatoes is now on the rise. Findings have shown that this type of film may prevent postharvest losses by reducing oxidative damage through effective regulation of the ascorbate–glutathione cycle. Li *et al.*[45] investigated the effect of preservative (rosmarinic acid coated with chitosan) in the presence of various concentrations of polylysine during storage at 4 °C on the organoleptic properties of half-smooth tongue sole (HTS) fillets. HTS fillets were coated with an average amount of off-smell organic compound solvents such as propanal, octanal, esters, 1-octen-3-ol and hexanal, which was greatly reduced. Sensory evaluation studies have shown that a mixture of polylysine and chitosan-based coating along with rosmarinic acid is an effective and reliable technique for keeping HTS fillets in good condition during refrigeration. Because of the large number of amine groups responsible for chitosan's cationic characteristic, it has a far stronger antibacterial impact than chitin.

In an acidic environment, positive and negative charges interact with residues of lipids, proteins and carbohydrates on the cell surface of bacteria, inhibiting their growth.[46,47] Takahashi and team used two distinct testing methods, mannitol salt agar medium incubation and also the use of the conductimetric assay to determine the rate at which deacetylation (DDA) of chitosan affected its antibacterial effectiveness against *Staphylococcus aureus*.[48] The results obtained indicated that the inhibition of *Staphylococcus aureus* was better achieved in both methods using DDA of chitosan. According to Younes *et al.*[49] antibacterial activity based on chitosan DDA produced comparable results. Chitosan was found to inhibit almost all species of bacteria used in the experiment at the lowest concentration when the DDA was about 100 percent (99 percent). The antibacterial ability of chitosan microparticles against *Klebsiella pneumoniae*, *Streptococcus uberis*, *E. coli* and *Salmonella enterica* was demonstrated by Jeon *et al.*[50]

The antibacterial potential of chitosan microparticles was observed to be effective against a large group of microorganisms, especially bacteria and fungi, and the inhibitory activity increased when a higher dose of chitosan microparticles was used. Despite the numerous researches conducted to date, there remains a failure in concluding a clear relationship between chitosan's antimicrobial properties and its molecular weight and DDA. This could be due to a variety of different factors affecting the inhibition rate, such as the bacterial strain types and biological testing settings.[51]

Xing *et al.*[52] investigated the preservation mechanism of chitosan-based coating with cinnamon oil for fruit storage based on the sensor data. Many micropores were discovered in the chitosan coating films using atomic force microscopy sensor images. The chitosan concentration influences the roughness of the coated film. Cinnamon oil's antifungal action is thought to be mostly due to its major constant *trans*-cinnamaldehyde, which is proportional to *trans*-cinnamaldehyde concentration and this is enhanced as the oil attachment time increases. Increasing the proportion of cinnamon oil could enhance the exosmosis ratios of *Penicillium citrinum* and *Aspergillus flavus*. The withered mycelium of *A. flavus* is detected surrounding the inhibition zone, compared to the normal cell, and the generation of spores is likewise inhibited, according to morphological observations.

Pabast *et al.*[53] used Satureja essential oil (SKEO)-loaded nanoliposomes to create a novel coating structure for lamb meat. Integrating a chitosan coating containing nano-encapsulated SKEO into lamb meat resulted in the preservation of high-quality attributes, improved microbiological quality and shelf-life extension during chilled storage, as well as improved oxidative stability. Previous studies have found that encapsulating essential oils in liposomes increased the antioxidant and antimicrobial activities compared to using them in their natural state.[54,55] As a result, the lower evaporation qualities and ease of transport to the bacterial cell wall may be explained. Plant extracts or phytochemicals utilized in the production of green-synthesized silver nanoparticles (AgNPs) may play important roles as alternatives with improved antimicrobial activities as compared to synthetic nanomaterials, according to Rezaei *et al.*[56]

Plant extracts in recent years have been established as suitable for nanomaterial production. Biogenic nanoparticles are products of plant extracts that contains amino acids and vitamins.[57] Plant extracts containing glucose or fructose are now frequently used for the formation of metal nano-based particles. *E. coli* extract with antifungal ability was used to synthesize gold nanoparticles which have been applied in protecting humans against infection.[58] *Delftia acidovorans* was discovered to be capable of synthesizing absolute gold nanocrystals by generating adequate resistance to harmful gold ions.

The outer capsid of viruses contains proteins with vast surface areas, which enables them to interact with several metallic ions, and this property makes them effective generators of nanoparticles.[59] Nanoparticles are currently created from waste and by-products, such as fruit (peels and seeds), coconut coir and palm oil, which contain significant levels of proteins, flavonoids and phenolic compounds, and they serve as reducing agents,[60,61] for example, the antimicrobial application of gold and silver nanoparticles obtained from fruit waste which inhibits foodborne diseases has been described.

Pesticides are known to be very permeable and soluble, are toxic in agronomy, and are commonly and heavily used. The analysis of highly sensitive pollutant residues is crucial.[62] This review chapter has discussed the successes and limitations described by many authors who have used nanoengineered sensors in food production and operations. Ranging from various processing methods and their preservation, packaging food products that are either processed or fresh, nanotechnology plays an important role when it comes to the selection of nanomaterials or maintaining food chains and their variety of functions, and also helps in the assurance of product quality and safety in the market.

References

1. R. Ravichandran, Nanotechnology applications in food and food processing: innovative green approaches, opportunities and uncertainties for global market, *Int. J. Green Nanotechnol.: Phys. Chem.*, 2010, **1**(2), P72–P96.
2. A. Fernando, K. D. Weerawardene, N. V. Karimova and C. M. Aikens, Quantum mechanical studies of large metal, metal oxide, and metal chalcogenide nanoparticles and clusters, *Chem. Rev.*, 2015, **115**(12), 6112–6216.
3. S. M. Alfadul and A. A. Elneshwy, USE of nanotechnology in food processing, packaging and safety – a review, *Afr. J. Food, Agric., Nutr. Dev.*, 2010, **10**(6), 2719–2739.
4. P. S. Aithal, Nanotechnology innovations & business opportunities: a review, *Int. J. Manage. IT Eng.*, 2016, **6**(1), 182–204.
5. B. Cuq, E. Rondet and J. Abecassis, Food powders engineering, between know-how and science: Constraints, stakes and opportunities, *Powder Technol.*, 2011, **208**(2), 244–251.
6. S. K. Ameta, A. K. Rai, D. Hiran, R. Ameta and S. C. Ameta, Use of Nanomaterials in Food Science, *Biog. NanoPart. Use Agroecosyst.*, 2020, 457–488.
7. X. He, H. Deng, W. G. Aker and H. M. Hwang, *Regulation and safety of nanotechnology in the food and agriculture industry, Food Applications of Nanotechnology*, Taylor & Francis Group, Boca Raton, 2019, vol. 12.
8. Cientifica Report, Nanotechnologies in the Food Industry, 2006. Available at: http://www.cientifica.com/www/details.php?id47 (accessed October 24, 2006).
9. F. Mustafa and S. Andreescu, Nanotechnology-based approaches for food sensing and packaging applications, *RSC Adv.*, 2020, **33**(10), 19309–19336.
10. X. Liu, M. Marrakchi, D. Xu, H. Dong and S. Andreescu, Biosensors based on modularly designed synthetic peptides for recognition, detection and live/dead differentiation of pathogenic bacteria, *Biosens. Bioelectron.*, 2016, **80**, 9–16.
11. X. Yan, H. Li and X. Su, Review of optical sensors for pesticides, *TrAC, Trends Anal. Chem.*, 2018, **103**, 1–20.

12. D. Wu, D. Du and Y. Lin, Recent progress on nanomaterial-based biosensors for veterinary drug residues in animal-derived food, *TrAC, Trends Anal. Chem.*, 2016, **83**, 95–101.
13. S. Y. Oh, M. J. Lee, N. S. Heo, S. Kim, J. S. Oh, Y. Lee, E. J. Jeon, H. Moon, H. S. Kim, T. J. Park and G. Moon, Cuvette-Type LSPR Sensor for Highly Sensitive Detection of Melamine in Infant Formulas, *Sensors*, 2019, **19**, 3839.
14. X. Lin, Y. Ni and S. Kokot, Glassy carbon electrodes modified with gold nanoparticles for the simultaneous determination of three food antioxidants, *Anal. Chim. Acta*, 2013, **765**, 54–62.
15. E. Sharpe, T. Frasco, D. Andreescu and S. Andreescu, Portable ceria nanoparticle-based assay for rapid detection of food antioxidants (NanoCerac), *Analyst*, 2013, **138**, 249–262.
16. S. H. Nile, V. Baskar, D. Selvaraj, A. Nile, J. Xiao and G. Kai, Nanotechnologies in Food Science: Applications, Recent Trends, and Future Perspectives, *Nano-Micro Lett.*, 2020, **12**, 45.
17. S. T. G. Gedarawatte, J. T. Ravensdale, H. Al-Salami, B. K. Ndwandwe, S. P. Malinga, E. Kayitesi and B. C. Dlamini, Advances in green synthesis of selenium nanoparticles and their application in food packaging, *Int. J. Food Sci. Technol.*, 2021, **56**, 2640–2650.
18. D. Deus, C. Kehrenberg, D. Schaudien, G. Klein and C. Krischek, Effect of a nano-silver coating on the quality of fresh turkey meat during storage after modified atmosphere or vacuum packaging, *Poult. Sci.*, 2017, **96**, 449–457.
19. Y. Huang, L. Mei, X. Chen and Q. Wang, Recent Developments in Food Packaging Based on Nanomaterials, *Nanomaterials*, 2018, **8**, 830.
20. M. Cushen, J. Kerry, M. Morris, M. Cruz-Romero and E. Cummins, Evaluation and simulation of silver and copper nanoparticle migration from polyethene nanocomposites to food and an associated exposure assessment, *J. Agric. Food Chem.*, 2014, **62**, 1403–1411.
21. Y. Huang, L. Mei, X. Chen and Q. Wang, Recent Developments in Food Packaging Based on Nanomaterials, *Nanomaterials*, 2018, **8**, 830.
22. A. Fernández, E. Soriano, G. López-Carballo, P. Picouet, E. Lloret, R. Gavara and P. Hernández-Muñoz, Preservation of aseptic conditions in absorbent pads by using silver nanotechnology, *Food Res. Int.*, 2009, **42**, 1105–1112.
23. A. Martínez-Abad, J. M. Lagaron and M. J. Ocio, Development and Characterization of Silver-Based Antimicrobial Ethylene–Vinyl Alcohol Copolymer (EVOH) Films for Food-Packaging Applications, *J. Agric. Food Chem.*, 2012, **60**, 5350–5359.
24. B. Panea, G. Ripoll, J. González, Á. Fernández-Cuello and P. Albertí, Effect of nanocomposite packaging containing different proportions of ZnO and Ag on chicken breast meat quality, *J. Food Eng.*, 2014, **123**, 104–112.
25. H. Yu, J. Y. Park, C. W. Kwon, S. C. Hong, K. M. Park and P. S. Chang, An overview of nanotechnology in food science: preparative methods, practical applications, and safety, *Journal of Chemistry*, 2018, 1–10.

26. B. Yalcin and S. Otles, Nano biosensor and food pathogen interaction mechanisms, *Electron. J. Environ., Agric. Food Chem.*, 2010, **9**(7), 1257–1273.
27. C. O. Adetunji, S. A. Akande, A. K. Oladipo, R. A. Salawu and A. F. Onyegbula, Determination of the microbiological quality and proximate composition of fermented cassava food products sold in Ilorin-west local government area, Nigeria, *Ruhuna J. Sci.*, 2017, **8**, 76–89.
28. F. F. Olayemi and C. O. Adetunji, Effect of rinses on microbial quality of commercially available eggs and their components before processing from Ilorin in western Nigeria, *Bitlis Eren Univ. J. Sci. Technol.*, 2013, **3**(2), 44–47.
29. C. O. Adetunji, O. B. Fawole, K. A. Arowora, S. I. Nwaubani, E. S. Ajayi, J. K. Oloke, J. A. Aina, J. B. Adetunji and A. O. Ajani, Postharvest quality and safety maintenance of the physical properties of Daucus *carota L.* fruits by Neem oil and Moringa oil treatment: A new edible coatings, *Agrosearch*, 2013, **13**(1), 131–141.
30. C. O. Adetunji, S. M. Olugbenga, K. Oseni, V. Ajit, A. Muhammad, K. O. Julius, S. Hamda, B. A. Juliana, J. Aditi, E. B. Ruth and O. Phebean, Benjamin Ewa Ubi Quinoa: From Farm to Traditional Healing, Food Application, and Phytopharmacology, in *Biology and Biotechnology of Quinoa*, ed. A. Varma, Springer, Singapore, 2021.
31. C. O. Adetunji, S. M. Olugbenga, W. Nwankwo, E. U. Kingsley, A. A. Osikemekha, K. O. Julius, V. Ajit, K. Oseni, J. Aditi and B. A. Q. Juliana, The Next Biotech Plant: Food Security and Environmental and Health Hot Spots, in *Biology and Biotechnology of Quinoa*, ed. A. Varma, Springer, Singapore, 2021.
32. C. O. Adetunji, S. M. Olugbenga, V. Ajit, K. Julius, K. Oseni, A. Muhammad, E. B. Ruth, I. Areeba, B. A. Juliana, S. Khuram, J. Aditi, E. U. Benjamin, M. Noshiza, O. Phebean and N. O. Frances, Recent Advances in the Application of Biotechnology for Improving the Production of Secondary Metabolites from Quinoa, in *Biology and Biotechnology of Quinoa*, ed. A. Varma, Springer, Singapore, 2021.
33. C. O. Adetunji, T. O. Olugbemi, A. Olorunsola, D. Ayobami, J. A. Mayowa, A. Enoch, R. Maksim, G. Larisa and A. S. Mohammad *et al.*, eHealth, mHealth, and Telemedicine for COVID-19 Pandemic, in *Assessing COVID-19 and Other Pandemics and Epidemics using Computational Modelling and Data Analysis*, ed. S. K. Pani, S. Dash, W. P. dos Santos, S. A. Chan Bukhari and F. Flammini, Springer, Cham, 2022.
34. O. A. Charles, T. O. Olugbemi, A. Olorunsola, D. Ayobami, J. A. Mayowa, A. Enoch, R. Maksim, P. Ekaterina and A. S. Mohammad, Machine Learning Approaches for COVID-19 Pandemic, in *Assessing COVID-19 and Other Pandemics and Epidemics using Computational Modelling and Data Analysis*, ed. S. K. Pani, S. Dash, W. P. dos Santos, S. A. Chan Bukhari and F. Flammini, Springer, Cham, 2022.
35. O. A. Charles, T. O. Olugbemi, A. Olorunsola, D. Ayobami, J. A. Mayowa, A. Enoch, R. Maksim, I. Olga and A. S. Mohammad, Smart Sensing for

COVID-19 Pandemic, in *Assessing COVID-19 and Other Pandemiing and Data Analysis*, ed. S. K. Pani, S. Dash, W. P. dos Santos, S. A. Chan Bukhari and F. Flammini, Springer, Cham, 2022, pp. 145–156.
36. O. A. Charles, T. O. Olugbemi, A. Olorunsola, D. Ayobami, J. A. Mayowa, A. Enoch, R. Maksim, P. Ekaterina and A. S. Mohammad, Internet of Health Things (IoHT) for COVID-19, in *Assessing COVID-19 and Other Pandemics and Epidemics using Computational Modelling and Data Analysis*, ed. S. K. Pani, S. Dash, W. P. dos Santos, S. A. Chan Bukhari and F. Flammini, Springer, Cham, 2022, pp. 75–87.
37. O. A. Charles, T. O. Olugbemi, A. Olorunsola, D. Ayobami, J. A. Mayowa, A. Enoch, R. Maksim, K. Natalia and A. S. Mohammad, Diverse Techniques Applied for Effective Diagnosis of COVID-19, in *Assessing COVID-19 and Other Pandemiing and Data Analysis*, ed. S. K. Pani, S. Dash, W. P. dos Santos, S. A. Chan Bukhari and F. Flammini, Springer, Cham, 2022, pp. 45–58.
38. F. Mustafa and S. Andreescu, Chemical and Biological Sensors for Food-Quality Monitoring and Smart Packaging, *Foods*, 2018, 7(10), 168.
39. P. Priya and D. Abdulgaffar, Role of Sensor In The Food Processing Industries, *Int. Arch. App. Sci. Technol*, 2019, **10**(1), 10–18.
40. C. Adley, Past, Present and Future of Sensors in Food Production, *Foods*, 2014, **3**(3), 491–510.
41. T. P. Anand, C. Chellaram, G. Murugaboopathi, V. Parthasarathy and S. Vinurajkumar, Applications of Biosensors in Food Industry, *Biosci., Biotechnol. Res. Asia*, 2013, **10**(2), 711–714.
42. B. D. Meshram, A. K. Agrawal, A. Shaikh, R. Suvartan and K. K. Sande, Biosensor and its application in food and dairy industry: A Review, *Int. J. Curr. Microbiol. Appl. Sci.*, 2018, 7(2), 3305–3324.
43. N. K. Murali, B. D. S. Srinivas and B. S. K. Jakeer, Biosensors in Food Processing – A Review, *Int. J. Pure App. Biosci.*, 2017, 5(4), 1219–1227.
44. J. Zhu, H. Wu and Q. Sun, Preparation of crosslinked active bilayer film based on chitosan and alginate for regulating ascorbate-glutathione cycle of postharvest cherry tomato (*Lycopersicon esculentum*), *Int. J. Biol. Macromol.*, 2019, **130**, 584–594.
45. N. Li, W. Liu, Y. Shen, J. Mei and J. Xie, Coating effects of ε-polylysine and rosmarinic acid combined with chitosan on the storage quality of fresh half-smooth tongue sole (Cynoglossus semilaevis Günther) fillets, *Coatings*, 2019, **9**(4), 273.
46. T. P. Anand, C. Chellaram, G. Murugaboopathi, V. Parthasarathy and S. Vinurajkumar, Applications of Biosensors in Food Industry, *Biosci., Biotechnol. Res. Asia*, 2013, **10**(2), 711–714.
47. J. Kumirska, M. X. Weinhold, J. Thöming and P. Stepnowski, Biomedical activity of chitin/chitosan-based materials—influence of physicochemical properties apart from molecular weight and degree of N-acetylation, *Polymer*, 2011, **3**(4), 1875–1901.

48. T. Takahashi, M. Imai, I. Suzuki and J. Sawai, Growth inhibitory effect on bacteria of chitosan membranes regulated with deacetylation degree, *Biochem. Eng. J.*, 2008, **4**(3), 485–491.
49. I. Younes, S. Sellimi, M. Rinaudo, K. Jellouli and M. Nasri, Influence of acetylation degree and molecular weight of homogeneous chitosans on antibacterial and antifungal activities, *Int. J. Food Microbiol.*, 2014, **185**, 57–63.
50. S. J. Jeon, M. Oh, W. S. Yeo, K. N. Galvão and K. C. Jeong, Underlying mechanism of antimicrobial activity of chitosan microparticles and implications for the treatment of infectious diseases, *PLoS One*, 2014, **9**(3), 1–7.
51. M. Kong, X. G. Chen, K. Xing and H. J. Park, Antimicrobial properties of chitosan and mode of action: a state-of-the-art review, *Int. J. Food Microbiol.*, 2010, **144**(1), 51–63.
52. Y. Xing, Q. Xu, S. X. Yang, C. Chen, Y. Tang, S. Sun, L. Zhang, Z. Che and X. Li, Preservation Mechanism of Chitosan-Based Coating with Cinnamon Oil for Fruits Storage Based on Sensor Data, *Sensors*, 2016, **16**(7), 1111.
53. M. Pabast, N. Shariatifar, S. Beikzadeh and G. Jahed, Effects of chitosan coatings incorporating with free or nano-encapsulated Satureja plant essential oil on quality characteristics of lamb meat, *Food Control*, 2018, **91**, 185–192.
54. M. Rai, P. Paralikar, P. Jogee, G. Agarkar, A. P. Ingle, M. Derita and S. Zacchino, Synergistic antimicrobial potential of essential oils in combination with nanoparticles: Emerging trends and future perspectives, *Int. J. Pharm.*, 2017, **519**, 67–78.
55. M. Ghaderi-Ghahfarokhi, M. Barzegar, M. A. Sahari, H. Ahmadi Gavlighi and F. Gardini, Chitosan-cinnamon essential oil nano-formulation: Application as a novel additive for controlled release and shelf- life extension of beef patties, *Int. J. Biol. Macromol.*, 2017, **102**, 19–28.
56. A. Rezaei, M. Fathi and S. M. Jafari, Nanoencapsulation of hydrophobic and low-soluble food bioactive compounds within different nanocarriers, *Food Hydrocolloids*, 2019, **88**, 146–162.
57. J. Singh, T. Dutta, K.-H. Kim, M. Rawat, P. Samddar and P. Kumar, Green' synthesis of metals and their oxide nanoparticles: Applications for environmental remediation, *J. Nanobiotechnology*, 2018, **16**, 84.
58. M. Gholami-Shabani, M. Shams-Ghahfarokhi, Z. Gholami-Shabani, A. Akbarzadeh, G. Riazi, S. Ajdari, A. Amani and M. Razzaghi-Abyaneh, Enzymatic synthesis of gold nanoparticles using sulfite reductase purified from Escherichia coli: A green eco-friendly approach, *Process Biochem.*, 2015, **50**, 1076–1085.
59. V. Makarov, A. Love, O. Sinitsyna, S. Makarova, I. Yaminsky, M. Taliansky and N. Kalinina, "Green" nanotechnologies: Synthesis of metal nanoparticles using plants, *Acta Nat.*, 2014, **6**, 35–44.
60. M. A. Ebrahimzadeh, A. Naghizadeh, O. Amiri, M. Shirzadi-Ahodashti and S. Mortazavi-Derazkola, Green and facile synthesis of Ag

nanoparticles using Crataegus pentagyna fruit extract (CP-AgNPs) for organic pollution dyes degradation and antibacterial application, *Bioorg. Chem.*, 2020, **94**, 103425.
61. S. Ali, X. Chen, M. Ajmal-Shah, M. Ali, M. Zareef, M. Arslan, S. Ahmad, T. Jiao, H. Li and Q. Chen, The avenue of fruit wastes to worth for synthesis of silver and gold nanoparticles and their antimicrobial application against foodborne pathogens: A review, *Food Chem.*, 2021, **359**, 129912.
62. A. Mclaren, T. Valdes-Solis, G. Li and S. C. Tsang, Shape and size effects of ZnO nanocrystals on photocatalytic activity, *J. Am. Chem. Soc.*, 2009, **131**(35), 12540–12541.

CHAPTER 8

Nanobiomaterials for Food Packaging Sensor Applications

CHARLES OLUWASEUN ADETUNJI,*[a]
FRANK ABIMBOLA OGUNDOLIE,[b]
OLUGBEMI TOPE OLANIYAN,[c] JOHN TSADO MATHEW,[d]
ABEL INOBEME,[e] OLOTU TITILAYO,[f] SHAKIRA GHAZANFAR,[g]
OLUWATOSIN ADEMOLA IJABADENIYI,[h]
MODUPE DORIS AJIBOYE,[i] OLUOPE OLUFEMI AJAYI,[j]
WADZANI PALNAM DAUDA[k] AND JULIANA BUNMI ADETUNJI[l]

[a] Applied Microbiology, Biotechnology and Nanotechnology Laboratory, Department of Microbiology, Edo University Iyamho, PMB 04, Auchi, Edo State, Nigeria; [b] Department of Biotechnology, Baze University Abuja, Nigeria; [c] Laboratory for Reproductive Biology and Developmental Programming, Department of Physiology, Rhema University Aba, Nigeria; [d] Department of Chemistry, Ibrahim Badamasi Babangida University Lapai, Niger State 9111014, Nigeria; [e] Department of Chemistry, Edo University Iyamho, Nigeria; [f] Department of Microbiology, Adeleke University, Ede Osun State, Nigeria; [g] Functional Genomics and Bioinformatics, National Agricultural Research Centre, Islamabad 45500, Pakistan; [h] Department of Biotechnology and Food Science, Durban University of Technology, Durban 4001, South Africa; [i] Department of Biochemistry, Edo University Iyamho, Edo State, Nigeria; [j] Department of Biochemistry, Edo State University Uzairue, formerly Edo University Iyamho; [k] Crop Science Unit, Department of Agronomy, Federal University Gashua, Gashua P.M.B 1005, Yobe State, Nigeria; [l] Department of Biochemistry, Osun State University, Nigeria
*Emails: adetunjicharles@gmail.com; adetunji.charles@edouniversity.edu.ng

Food Chemistry, Function and Analysis
Bio- and Nano-sensing Technologies for Food Processing and Packaging
Edited by Ashutosh Kumar Shukla

Published by the Royal Society of Chemistry, www.rsc.org

8.1 Introduction

Bionanotechnology is becoming increasingly important in the realm of food packaging as a method for incorporating biological materials into nanoartefacts. It provides a higher level of food packaging standard, ensuring prolonged shelf life of the product as well as safer packaging with increased quality of food and accountability. Recent improvements in food nanopackaging, such as bio-based, enhanced, active, and also intelligent packing, are the subject of this chapter. Bio-based packaging, particularly biocompatible and biodegradable packaging, is emphasized as more than just a viable alternative to one of the most often utilized non-degradable polymeric composites.

Nanocomposites and nanoparticles have useful application in food packaging due to the antimicrobial properties of these materials. Issues have been expressed concerning nanotechnology transmission from packaging to food, although the results of migratory testing, as well as risk assessments, are ambiguous. The utilization of nanoparticles in packaged foods is limited due to suspected cytotoxicity, and a shortage of adequate information from risk assessment studies and clinical studies.[1]

Nanotechnology is a recently developed revolutionary food packaging technique that could extend the shelf life of foods, maintain food safety, eliminate spoilage, alleviate food shortages, and ultimately enhance people's health as well as avoiding packaging tears. It has the ability to release inhibitors in order to enhance the shelf life of the food in the package.

Packaged foods, as well as surveillance, are presently a prominent focus of nanoscience and nanotechnology in the food sector. Nanotechnology has a great deal of potential in the food industry, as it could be utilized to in the manufacture of around 25 percent of all packaged foods in the coming years. It remains unknown if nanotechnology utilized during food packaging can move or leak into food. The possible health effects of nanomaterials should be thoroughly investigated. Ensuring the safety of chemicals in nanotechnology is prioritized before they are used in packaged foods, particularly in the packaging.[2]

Nanotechnology is an innovative frontier mostly in the transformation of traditional agricultural production as well as the food business into a unique concept for food industrial growth. Nanosensors, nanofoods, nanofertilizers, nanopesticides as well as nanopackaging are some of the most prominent nanotechnologies and also the most significant technological breakthroughs. Food safety, food quality, and packaged food features such as nanonutraceuticals, nanofood and drug delivery, and food supplements are all impacted by nanoscience-based technologies. Nutrition enhancement, food preservation, especially bioactive components and safe distribution of micronutrients are all made easier using nanomaterials.[3]

This recent research makes available the developments and advancements using nanotechnology, as well as revealing its intriguing possibilities and problems in the food processing business. In this chapter, the production of nanomaterials with its potential utilizations in the food industry are discussed, as well as health regulation and hazard analysis challenges. Despite the fact that nanotechnology is a viable potential technology with advanced applications in the food business, more initiatives are required to conduct extensive research into the nanofood scheme and to increase awareness among consumers.[4]

Nanotechnology is a new frontier that is converting the local food industry into one that is dynamic, innovative and emergent. The rapid evolution of nanotechnology has already accelerated changes to traditional food, particularly the construction of stylish and vivid packaging, as well as the development of innovative nanomaterials to improve food safety and quality. Among the most current advancements in nanoscience are nanobiosensors, nanofood as well as nanopackaging. Nanoscience-based research appears to have a significant effect on the safety and quality in the packaging of food components.

Biofortification of food, storage period extension, including protein enhancement, are all possible thanks to nanotechnology. Despite the numerous advantages of nanotechnology, there remains serious concerns about its application, as the concentration of nanoparticles in human health and the environment could pose a variety of health and safety risks. The fundamentals of toxicology and risk assessment of nanoparticles in innovative diets are also discussed. The potential applications of biosynthesized and bioinspired nanomaterials are highlighted for equitable and comprehensive development.

However, critical questions about better performing, less-noxious nanomaterials must be addressed in order to permit progressive growth and application of nanotechnology. Regulation and legislation are critical for managing the use, manufacture, treatment, handling and disposal of nanoparticles. This will take a great deal of effort in increasing awareness and acceptance of novel nano-enabled meals.[5]

Some of the rapid evolutions in nanotechnology have also paved the way for the creation of food packaging solutions and innovative sensing, resolving long-standing difficulties in the food industry such as extending shelf life, reducing waste, determining the safety of food, and improving its quality. Packaged foods could benefit from nanomaterials to improve gas barrier qualities, mechanical stability, boost water repellency, and impart antibacterial as well as oxidizing action. These could be used in biological and chemical sensors to create fast and sensitive sensors for determining freshness, including detecting allergies, poisons, and infectious pollutants. This chapter discusses current research on the utilization of nanotechnology in the advancement of (1) (bio)sensors for detecting micronutrient and non-nutritive components, adulterants, toxicants, and antioxidants, (2) active functional packaging, and (3) techniques to enhance the mechanical and barrier properties of food packaging. Nanomaterials in the food industry have health, safety, and environmental consequences, as well as requiring an assessment of legislation and customer perception.[6,7]

The use of nanotechnology through packaged foods has been widely discussed over the last 20 years. Throughout this context, several areas of research on the use of nanotechnology, including clay, zinc oxide, carbon nanotube, silver, copper, titanium dioxide, along with copper oxide through packaged foods have already been undertaken, and are included in the current study. The use of nanopackaged foods improves the food's general physicochemical qualities (flavour, colour, weight, moisture content, texture, as well as bioavailability) and decreases the microbial consignment through Trojan-horse, reactive oxygen mechanisms and cell-membrane function, at the same time as advancing the mechanical/barrier properties of packaged food. Although nanopackaged foods provide several benefits for public health, the risk associated with migration, particularly in acidic environments, is significant. More research into the benefits and drawbacks of such a method is required.[8] Therefore, this chapter intends to provide detailed information on the application of nanobiomaterials for food packaging sensors.

8.2 The Application of Nanomaterials for Packaging of Foods and Relevance of Sensors

In recent years, nanomaterials have made a significant contribution to the advancement of food packaging technologies. Nanomaterials are materials with one or more dimensions between 1 and 100 nanometres. They are employed in food packaging for their optical, magnetic, electrical, and other properties. Some of the most common nanomaterials used for food packaging and preservation are nanoparticles (NPs), nanoclays (NCs), nanocomposites (NCs), nanostructures (NSTs) nanosensors (NSs), and nanoemulsions (NEs). The use of edible coatings, smart and active packaging, and the growth of a diverse variety of nanosystems offers answers to innovative food preservation and packaging. This makes nanosystems vital tools and solutions for controlling and increasing food safety and quality, which is greatly desired in food technology.[9]

Novel nanomaterials can even check the quality of food in real time, making them a viable option for food preservation. Nanomaterials in food carriers possess toxicological risk, especially in edible nanopackaging, which is strongly linked to the emergence of harmful consequences on the human body.[10–20] Nanocomposites are solid hybrid materials of numerous phases, having at least one dimension and phase in the nanoscale, with nanoparticles accounting for up to 5% of the total weight. Nanocomposites have antimicrobial activity against bacteria, yeast, and filamentous fungi,[21,22] has and have lower permeability in O_2 and CO_2.[23] Nanosensors embedded in packaging can detect all changes in food processing, which could be physical, chemical, and biological changes, that occur during the processing of food. Furthermore, smart packaging can detect poisons, harmful microorganisms, and chemicals using specialized nanosensors and nanotechnologies. To detect mycotoxins in foods, a variety of biosensors have been developed. A magnetic nano-gold immunosensor may detect aflatoxins produced by *Aspergillus parasiticus* and *Aspergillus flavus* in contaminated food. Also, pathogens like *Listeria monocytogenes*, *Salmonella*, and *E. coli* are detected using several nanobiosensors in smart packaging.[24]

The term "active packaging" refers to a system in which the food packaging interacts with the food and the inner part of the packaging in order to maintain the food's quality. In active food packaging, the release antioxidants and antibacterial chemicals into the packaging enhances food preservation more than in conventional food packaging, as well as capturing oxygen and water vapour, both of which are elements that degrade quality.[25] Nanoclay, nano-zinc oxide (nano-ZnO), silver NPs (AgNPs),[26] titanium nitride NPs (nano-TiN), nano-titanium dioxide (nano-TiO_2), and other engineered nanomaterials are currently being used as functional additions in food packaging.[27] Each nanomaterial has various properties for the host material due to differences in features and chemical structure, resulting in a variety of packaging applications. Many nanomaterials, such as titanium dioxide and zinc oxide, are frequently utilized as photocatalysts to disintegrate organic molecules and bacteria, while antimicrobial agents include nanoclays, layered silicates, and AgNPs.[28]

ZnO-NPs come in a variety of shapes and prevent the growth of a wide range of bacteria.[29] Cosmetics, medical equipment, pharmaceutical delivery, and modified atmospheric processed (MAP) packaging all contain zinc oxide nanoparticles (ZnONPs). ZnO nanostructures have been shown to be more effective at killing *Bacillus atrophaeus*, *E. coli*, and *Salmonella aureus* than other metal oxides.[30] ZnO-NPs are particularly appealing for packaging applications as compared to AgNPs because they are less expensive and less hazardous to animals and people.[31] TiO_2-NPs are metal oxide nanomaterials with thermo-stability, inertia, and are highly researched materials. This substance can also change the properties of biodegradable films, making them less expensive, nontoxic, and photo-stable. Food packaging has commonly used TiO_2 as a photocatalyst that has lots of economic and environmental applications such as water filtering, antibacterial, self-cleaning structures, and water and air purification.[32] Rosa *et al.*[33] generated cellulose nanowhiskers from delignified coconut husk fibre using a hydrolysis process using micro/nano-sized bamboo fibres to build a soy protein-based green composite that improved the Young's modulus and fracture stress significantly.[34]

Nanocellulose appears to be a possible replacement for standard petroleum-based polymers due to its exceptional characteristics and remarkable biodegradability. Antibacterial activity of zinc oxide particles is good, and improves as the particle size gets smaller, with stimulation; ZnO needs visible light, as described by Jones *et al.*[35] Microbial cell walls with direct interaction may result in bacterial cell death, the release of antibiotic ions such as Zn ions, and the production of reactive oxygen species, as described by Sirelkhatim *et al.*[36] Sawai[37] and Sevinc and Hanley[38] observed that when comparing MgO and CaO as antibacterial agents for *S. aureus* growth inhibition, ZnO was the most effective.

ZnO inhibits the growth of the bacteria biofilms of *Streptococcus sobrinus* in dental materials by roughly 80%. Jin *et al.*[39] investigated various methods for the application of nano-ZnO in food systems, including PVP capped, coating, powder, and film, which determined that nano-ZnO exhibited antibacterial activity against *L. monocytogenes* and *Salmonella enteritidis* in culture media and liquid egg medium. Antimicrobial activity in ZnO and ZnO powder nanoparticles was examined by Li *et al.*[40] against pathogenic microbes (*S. aureus*, *E. coli*, *Bacillus cereus*, *S. enteritidis*) and it was revealed that greater antibacterial activity was observed in ZnO nanoparticles than in the non-nanopowder against every bacterium tested against. Thymol film and polybutylene succinate were tested for antibacterial activity against *S. aureus* and *E. coli*.[41]

Nanosensors are nanotechnology-assisted sensors that come in an array of sizes and shapes, with the primary purpose of gas sensors being to show the presence of a gaseous analyte in a package. Hydrogen sulphide, volatile substances like amines, and CO_2 quality checks are enabled using opto-chemical sensors, which work on the principle of luminescence quenching or the absorbance changes caused by direct contact with the analyte.[42]

Electro-optical and chemical properties in nanoscale particles are affected by environmental variations such as the degree of oxygen exposure, temperature, and humidity, which in turn affect both product deterioration and microbial contamination.[43] Antibodies attached to nanomaterials like quantum dots (QDs) are primarily being investigated for bacterial detection. Owing to their excellent fluorescence efficiency, extended decay lifetime, high sensitivity, photobleaching stability, and electronic features having continuous and large narrow emission spectra and optical absorption, QDs are widely used.[44] *Salmonella typhimurium* and *E. coli* O157:H7 were detected using QDs, according to Yang and Li.[45] Anti-*E. coli* O157 and anti-*Salmonella* antibodies were combined with highly fluorescent CdSe/ZnS QDs. The emission wavelengths of QDs for *Salmonella* and *E. coli* were different, but the excitation wavelengths were the same, which allows for the detection of the two test pathogens simultaneously. In comparison to traditional fluorescent dyes, the QD conjugate technique has various advantages.

QDs are resistant to photobleaching, can be detected numerous times using various QDs, and have substantially higher fluorescence intensity.[46] Rotting of meat was detected using a coating plastic film with a transition metal layer of 1–10 nm thickness when sulphide volatiles were created by spoiling fresh meats that react with this layer, turning it black.[47] Maynor *et al.*[48] used conjugated polythiophene to create multidimensional colorimetric effects that accurately discriminated 22 structurally similar amines. For the identification of volatile amines in tuna, the conjugated system was used. Later, a polyaniline film was produced that was sensitive to volatile amines generated during fish deterioration.

These interventions can help detect fish deterioration in real time at various temperatures, as reported by Kuswandi *et al.*[49] A portable bioelectronic nose based on peptide receptors for the detection of food freshness detecting trimethylamine, a major volatile generated during raw seafood rotting, by Lim *et al.*[50] Carbon nanotubes have been used to create a sensor that can detect volatile chemicals, ethylene, and CO_2 released during fruit ripening.[51]

Increased food productivity with cost-effective returns, innovative goods with tuneable qualities to produce smarter and healthier foods, and intelligent packaging solutions with enhanced storage properties for better food safety would benefit nations throughout the world. With the emergence of novel nanomaterials for application in the food sector, however, the task of determining the safety of nanofoods and nanopackaging has become increasingly complex. As a result, greater coordination is required to ensure that social and environmental interests are not jeopardized as new goods are launched. Finally, the rate at which food technology is introduced must be reduced sufficiently to allow for the identification and assessment of potential concerns in order to ensure a safer future.

Chaudhary *et al.*[52] in their study reported that biopolymers together with nanomaterials have been useful in the area of food packaging for the purpose of preventing hazards due to the nontoxicity and biodegradability of these materials. They highlighted the merits of using these when incorporated with nanomaterials. They were of the opinion that the unique properties of nanomaterials have made their application in food packaging possible and more promising. They also outlined some of the nanomaterials that have been employed in the area of food packaging which include zinc oxide, silver nanoparticles, coated silicates, and titanium oxides amongst others. They identified the compatibility and complete distribution of nanocomposites within the polymeric matrix as the major challenge.

Similarly, Sharma *et al.*[53] in a related work reported that the primary purpose of packaging and hygienically transporting food commodities is to ensure their preservation from undesired changes in quality prior to getting to the final consumer. They further observed that there has been a gradual evolution in the area of food packaging with advancements and innovations in science and nanotechnology. According to these researchers, the introduction of nanotechnology in the packaging of food has remarkably addressed the issues of food safety and quality concerns. In their review work, they focused on the current knowledge in this regard, current advances in the field of packaging as well as the envisaged future trends in the utilization of nanotechnology in the packaging of food.

Agriopouou *et al.*[9] highlighted the role of nanomaterials in the development of packaging systems of food within recent years. They opined that nanomaterials are capable of providing reliable solutions in food preservation and packaging through smart and active packaging systems. They also observed that nanomaterials could provide an option that is highly efficient in limiting, controlling, and improving the safety parameters of the quality of food. They also concluded that innovative nanoparticles are able to achieve real-time monitoring of food quality, which is paramount in food preservation.

Kumar *et al.*[54] observed that one of the major challenges for developing nations is the safe use and sustainable utilization of various agricultural products. They, therefore opined that the preservation of food materials without losing their nutritional values, shelf life, and freshness is paramount in boosting agricultural productivity. They, went on to highlight the various applications of nanomaterials in different areas such as nanobiosensors, medical applications, as well as usage in food packaging. Reports from various studies in this regard were also presented in their work.

Mustapha and Andreescu[7] opined that the rapid improvements in the area of nanotechnology have given rise to opportunities for the fabrication and development of novel food packaging and sensing solutions, reduction of waste, improvement of food quality, and extension of the shelf life of food. They also described some of the remarkable applications of nanotechnology in the food sector. They reported works based on the incorporation of nanomaterials in biological and chemical sensing systems for the quick fabrication of sensitive and rapid testing of allergens, pathogenic contaminants, and others. They reported recent investigations into the application of nonmaterial for the development of sensing technologies suitable for the detection of various substances.

Duncan,[55] in his review work, discussed the various applications of nanomaterials in the areas of food safety and packaging. He highlighted the use of polymeric nanocomposites as a barrier packaging material. His review work further addressed the most recent state of knowledge related to the health impacts of this technology. The role of silver nanoparticles in combating microorganisms and assays based on nanomaterials in the detection of various food-based analytes was further discussed.

Shafiq *et al.*[56] opined that the present advancements in nanotechnology have led to the transformation of various aspects of the food industry and food science with improved market share and general investment. They identified various areas of advancement in food science in connection with food packaging and nanotechnology which include nanoemulsions, nanosensors, nanocapsules, and nanopesticides, with the primary intention of bringing about innovative utilization in the food industry. They explored the most recent applications of nanoscience and technology in the processing, packaging, and improvement of the nutritional quality and shelf life of food products.

Primozic *et al.*[1] emphasized nanotechnology as an indispensable tool vital for the incorporation of biomolecules into various food materials as well as nanoartefacts, hence gaining more applications in the food industry. They positioned that nanotechnology provides an improved expectation of food product packaging which can ensure better quality and shelf life of products. Sundramoorthy and Gunasekaran[57] outlined the remarkable properties of various graphene-based nanomaterials and their derivatives such as electrical, mechanical, optical, thermal, and absorptive. They provided a detailed state-of-the-art evaluation of the use of graphene in the food sector. They examined critically the development and advancement in the synthesis of graphene, their uses in food analyses, as well as analytical techniques that are based on graphene and their applications in the food industries. They also discussed the antimicrobial properties of nanomaterials that are graphene based in food packaging.

Noor *et al.*[58] identified various safety issues that are vital for the formulation and development of nanomaterials that are applied in packaging systems. They observed that it is paramount to form and advance some approaches for the detection and determination of nanomaterials, especially when applied *in vivo*. They identified that nanomaterials are remarkably promising for food sector applications, as a result of their unique sizes and physicochemical properties, and hence their roles have been extended for enhancing the bioavailability and stability of the nutrients. Their review work focused on the packaging of food together with bio-based enhanced smart packaging. They placed unique emphasis on biological-based food packaging together with biodegradable and biocompatible packaging, which is an alternative to the most commonly utilized polymeric materials that are non-biodegradable.

Siddiqui *et al.*[59] highlighted that the wastage of food is a pressing issue of global concern. They identified the need for monitoring of the quality of food through the use of real-time analysis so as to meet global standards and hence the embedding of sensors containing active nanoparticles within packaged food would enable customers to monitor the quality of the food substances.

8.3 Conclusion

The future of food safety concerns in the various sectors of the food industries ranging from dairy, meat/pork, and mainstream food industries could be well addressed through smart packaging which involves the use of biosensors in conventional packaging materials. With the use of biosensor-based nanocomposite materials replacing conventional packaging materials, this technology is rapidly opening up the food sector into an era where packaging, detecting micronutrient components, toxicants, fibres, adulterants, and antioxidants can be well detected and monitored, thus ensuring food reliability, safety, and accountability, and reducing foodborne illness as a result.

References

1. M. Primozic, Z. Knez and M. Leigeb, (Bio)Nanotechnology in Food Science—Food Packaging, *Nanomaterials*, 2021, **11**(2), 292.
2. M. Pal, Nanotechnology: A New Approach in Food Packaging, *J. Food Microbiol. Saf. Hyg.*, 2017, **2**(2), DOI: 10.4172/2476-2059.1000121.
3. A. Sridhar, M. Ponnuchamy, P. S. Kumar and A. Kapoor, Food preservation techniques and nanotechnology for increased shelf life of fruits, vegetables, beverages and spices: a review, *Environ. Chem. Lett.*, 2020, **19**, 1715–1735.

4. T. B. Rouf and J. L. Kokini, Natural biopolymer-based nanocomposite films for packaging applications, *Bionanocomposites for Packaging Applications*, Springer, New York, NY, USA, 2018, pp. 149–177.
5. S. Siddiqui and S. A. Alrumman, Influence of nanoparticles on food: An analytical assessment, *J. King Saud Univ., Sci.*, 2021, **33**(6), 101530.
6. A. Lavado-Cruz, J. Huamán-Cano and L. M. Paucar-Menacho, Nanotechnology in the development of food packaging: A super technology that successfully addresses the current challenges of packaging and friendly to the environment, *Agroind. Sci.*, 2020, **10**(2), 211–217.
7. F. Mustapha and S. Andreescu, Nanotechnology-based approaches for food sensing and packaging applications, *RSC Adv.*, 2020, **10**(33), 19309–19336.
8. M. A. Emamhadi, M. Sarafraz, M. Akbari, V. N. Thai, Y. Fakhri, N. T. T. Linh and A. M. Khaneghah, Nanomaterials for food packaging applications: A systematic review, *Food Chem. Toxicol.*, 2020, **12**(2), 1–45.
9. S. Agriopoulou, E. Stamatelopoulou, V. Skiada, P. Tsarouhas and T. Varzakas, Emerging Nanomaterial Applications for Food Packaging and Preservation. Safety issues and Risk Assessment, *Proceedings*, 2020, **70**, 7.
10. C. O. Adetunji, S. A. Akande, A. K. Oladipo, R. A. Salawu and A. F. Onyegbula, Determination of the microbiological quality and proximate composition of fermented cassava food products sold in Ilorin-west local government area, Nigeria, *Ruhuna j. Sci.*, 2017, **8**, 76–89.
11. F. F. Olayemi and C. O. Adetunji, Effect of rinses on microbial quality of commercially available eggs and their components before processing from Ilorin in western Nigeria, *Bitlis Eren Univ. J. Sci. Technol.*, 2013, **3**(2), 44–47.
12. C. O. Adetunji, O. B. Fawole, K. A. Arowora, S. I. Nwaubani, E. S. Ajayi, J. K. Oloke, J. A. Aina, J. B. Adetunji and A. O. Ajani, Postharvest quality and safety maintenance of the physical properties of *Daucus carota L.* fruits by Neem oil and Moringa oil treatment: A new edible coatings, *Agrosearch*, 2013, **13**(1), 131–141.
13. C. O. Adetunji, S. M. Olugbenga, K. Oseni, V. Ajit, A. Muhammad, K. O. Julius, S. Hamda, B. A. Juliana, J. Aditi, E. B. Ruth and O. Phebean, Benjamin Ewa Ubi Quinoa: From Farm to Traditional Healing, Food Application, and Phytopharmacology, in *Biology and Biotechnology of Quinoa*, ed. A. Varma, Springer, Singapore, 2021.
14. C. O. Adetunji, S. M. Olugbenga, N. Wilson, E. U. Kingsley, A. A. Osikemekha, K. O. Julius, V. Ajit, K. Oseni, J. Aditi and B. A. Q. Juliana, The Next Biotech Plant: Food Security and Environmental and Health Hot Spots, in *Biology and Biotechnology of Quinoa*, ed. A. Varma, Springer, Singapore, 2021.
15. C. O. Adetunji, S. M. Olugbenga, V. Ajit, K. O. Julius, K. Oseni, A. Muhammad, E. B. Ruth, I. Areeba, B. A. Juliana, S. Khuram, J. Aditi, E. U. Benjamin, M. Noshiza, O. Phebean and N. O. Frances,

Recent Advances in the Application of Biotechnology for Improving the Production of Secondary Metabolites from Quinoa, in *Biology and Biotechnology of Quinoa*, ed. A. Varma, Springer, Singapore, 2021.
16. C. O. Adetunji, T. O. Olugbemi, A. Olorunsola, D. Ayobami, J. A. Mayowa, A. Enoch, R. Maksim, G. Larisa and A. S. Mohammad *et al.*, eHealth, mHealth, and Telemedicine for COVID-19 Pandemic, in *Assessing COVID-19 and Other Pandemics and Epidemics using Computational Modelling and Data Analysis*, ed. S. K. Pani, S. Dash, W. P. dos Santos, S. A. Chan Bukhari and F. Flammini, Springer, Cham, 2022.
17. O. A. Charles, T. O. Olugbemi, A. Olorunsola, D. Ayobami, J. A. Mayowa, A. Enoch, R. Maksim, P. Ekaterina and A. S. Mohammad, Machine Learning Approaches for COVID-19 Pandemic, in *Assessing COVID-19 and Other Pandemics and Epidemics using Computational Modelling and Data Analysis*, ed. S. K. Pani, S. Dash, W. P. dos Santos, S. A. Chan Bukhari and F. Flammini, Springer, Cham, 2022.
18. O. A. Charles, T. O. Olugbemi, A. Olorunsola, D. Ayobami, J. A. Mayowa, A. Enoch, R. Maksim, I. Olga and A. S. Mohammad, Smart Sensing for COVID-19 Pandemic, in *Assessing COVID-19 and Other Pandemiing and Data Analysis*, ed. S. K. Pani, S. Dash, W. P. dos Santos, S. A. Chan Bukhari and F. Flammini, Springer, Cham, 2022, pp. 145–156.
19. O. A. Charles, T. O. Olugbemi, A. Olorunsola, D. Ayobami, J. A. Mayowa, A. Enoch, R. Maksim, P. Ekaterina and A. S. Mohammad, Internet of Health Things (IoHT) for COVID-19, in *Assessing COVID-19 and Other Pandemics and Epidemics using Computational Modelling and Data Analysis*, ed. S. K. Pani, S. Dash, W. P. dos Santos, S. A. Chan Bukhari and F. Flammini, Springer, Cham, 2022, pp. 75–87.
20. C. O. Adetunji, A. Inobeme, J. Tadso, O. T. Olaniyan, F. A. Ogundolie, M. Shahnawaz and O. Anani, Potential of Plastic Waste in Enhancing the level of Pathogenicity of diverse Pathogens in the Marine Biota, *Impact of Plastic Waste on the Marine Biota*, Springer, Singapore, 2022, pp. 301–312.
21. Y. Pan, X. Huang, X. Shi, Y. Zhan, G. Fan, S. Pan, J. Tian, H. Deng and Y. Du, Antimicrobial application of nanofibrous mats self-assembled with quaternized chitosan and soy protein isolate, *Carbohydr. Polym.*, 2015, **133**, 229–235.
22. K. Hosseinnejad and S. M. Jafari, Evaluation of different factors affecting antimicrobial properties of chitosan, *Int. J. Biol. Macromol.*, 2016, **85**, 467–475.
23. T. B. Rouf and J. L. Kokini, Natural biopolymer-based nanocomposite films for packaging applications, in *Bionanocomposites for Packaging Applications*, Springer, New York, NY, USA, 2018, pp. 149–177.
24. K. Pathakoti, M. Manubolu and H. M. Hwang, Nanostructures: Current uses and future applications in food science, *J. Food Drug Anal.*, 2017, **25**, 245–253.
25. V. K. Bajpai, M. Kamle, S. Shukla, D. K. Mahato, P. Chandra, S. K. Hwang, P. Kumar, Y. S. Huh and Y. K. Han, Prospects of using

nanotechnology for food preservation, safety, and security, *J. Food Drug Anal.*, 2018, **26**, 1201–1214.
26. T. T. Hirvikorpi, M. Vähä-Nissi, T. Mustonen and M. Karppinen, Atomic layer deposited aluminium oxide barrier coatings for packaging materials, *Thin Solid Films*, 2010, **518**, 2654–2658.
27. *Packaging nanotechnology*, ed. A. K. Mohanty, M. Misra and H. S. Nalwa, American Scientific Publishers, 2009.
28. K. Majeed, M. Jawaid, A. Hassan, H. P. Abu Bakar, S. AKhalil, A. A. Salema and I. Inuwa, Potential materials for food packaging from nanoclay/natural fibres filled hybrid composites, *Mater. Des.*, 2013, **46**, 391–410.
29. A. Azam, A. S. Ahmed, M. Oves, M. S. Khan, S. S. Habib and A. Memic, Antimicrobial activity of metal oxide NPs against Gram-positive and Gram-negative bacteria: a comparative study, *Int. J. Nanomed.*, 2012, **7**, 6003–6009.
30. L. E. Shi, Z. H. Li, W. Zheng, Y. F. Zhao, Y. F. Jin and Z. X. Tang, Synthesis, antibacterial activity, antibacterial mechanism and food applications of ZnO NPs: a review, *Food Addit. Contam., Part A*, 2014, **31**(2), 173–186.
31. C. Silvestre, D. Duraccio and S. Cimmino, Food packaging based on polymer nanomaterials, *Prog. Polym. Sci.*, 2011, **36**, 1766–1782.
32. P. V. Kamat, Graphene-based nano assemblies for energy conversion, *J. Phys. Chem. Lett.*, 2011, **2**, 242–251.
33. M. F. Rosa, E. S. Medeiros, J. A. Malmonge, K. S. Gregorski, D. F. Wood, L. Mattoso, C. H. G. Glenn, W. J. Orts and S. H. Imam, Cellulose nanowhiskers from coconut husk fibers: effect of preparation conditions on their thermal and morphological behaviour, *Carbohyr. Polym.*, 2010, **81**, 83–92.
34. X. S. Huang and A. Netravali, Biodegradable green composites made using bamboo micro/nano-fibrils and chemically modified soy protein resin, *Compos. Sci. Technol.*, 2009, **69**, 1009–1015.
35. N. Jones, B. Ray, K. T. Ranjit and A. C. Manna, Antibacterial activity of ZnO nanoparticle suspensions on a broad spectrum of microorganisms, *FEMS Microbiol. Lett.*, 2008, **279**, 71–76.
36. A. Sirelkhatim, S. Mahmud, A. Seeni, N. H. Kaus, L. C. Ann, S. K. Bakhori, H. Hasan and D. Mohamad, Review on zinc oxide nanoparticles: antibacterial activity and toxicity mechanism, *Nano-Micro Lett.*, 2015 Jul, 7(3), 219–242.
37. J. Sawai, Quantitative evaluation of antibacterial activities of metallic oxide powders (ZnO, MgO and CaO) by conductimetric assay, *J. Microbiol. Meth.*, 2003, **54**, 177–182.
38. B. A. Sevinc and L. Hanley, Antibacterial activity of dental composites containing zinc oxide nanoparticles, *J. Biomed. Mater. Res., Part B*, 2010, **94**, 22–31.
39. T. Jin, D. Sun, J. Y. Su, H. Zhang and H. J. Sue, Antimicrobial efficacy of zinc oxide quantum dots against *Listeria monocytogenes, Salmonella enteritidis*, and *Escherichia coli* O157:H7, *J. Food Sci.*, 2009, **74**, 46–52.

40. X. Li, Y. Xing, Y. Jiang, Y. Ding and W. Li, Antimicrobial activities of ZnO powder-coated PVC film to inactivate food pathogens, *Int. J. Food Sci. Technol.*, 2009, **44**, 2161–2168.
41. N. Petchwattana and P. Naknaen, Utilization of thymol as an antimicrobial agent for biodegradable poly (butylene succinate), *Mater. Chem. Phys.*, 2015, **163**, 369–375.
42. K. B. Biji, C. N. Ravishankar, C. O. Mohan and T. K. Srinivasa Gopal, Smart packaging systems for food applications: a review, *J. Food Sci. Technol.*, 2015, **52**(10), 6125–6135.
43. H. Bouwmeester, S. Dekkers, M. Noordam, W. Hagens, A. Bulder, C. De Heer, S. E. ten Voorde, S. Wijnhoven and A. Sips, Health impact of nanotechnologies in food production. RIKILT Report 2007.014, RIVM rapport 000200702. 2007 Sep 1. Available online at: http://edepot.wur.nl/120669.
44. A. Valizadeh, H. Mikaeili, M. Samiei, S. M. Farkhani, N. Zarghami, A. Akbarzadeh and S. Davaran, Quantum dots: synthesis, bioapplications, and toxicity, *Nanoscale Res. Lett.*, 2012, 7(1), 1–4.
45. L. Yang and Y. Li, Simultaneous detection of *Escherichia coli* O157:H7 and *Salmonella Typhimurium* using quantum dots as fluorescence labels, *Analyst*, 2006, **131**, 394–401.
46. S. D. F. Mihindukulasuriya and L. T. Lim, Nanotechnology development in food packaging: A review, *Trends Food Sci. Technol.*, 2014, **40**(2), 149–167.
47. J. Smolander, K. Kuklane, D. Gavhed, H. Nilsson and I. Holmér, Effectiveness of a light-weight ice-vest for body cooling while wearing fire fighter's protective clothing in the heat, *Int. J. Occup. Saf. Ergon.*, 2004, **10**(2), 111–117.
48. M. S. Maynor, T. L. Nelson, C. O'Sulliva and J. J. Lavigne, A food freshness sensor using the multistate response from analyte-induced aggregation of a cross-reactive poly (thiophene), *Org. Lett.*, 2007, **9**(17), 3217–3220.
49. B. Kuswandi, A. Jayus Restanty, A. Abdullah, L. Y. Heng and M. Ahmad, A novel colorimetric food package label for fish spoilage based on polyaniline film, *Food Control*, 2012, **25**, 184–189.
50. J. H. Lim, J. Park, J. H. Ahn, H. J. Jin, S. Hong and T. H. Park, A peptide receptor-based bioelectronic nose for the real-time determination of seafood quality, *Biosens. Bioelectron.*, 2013, **39**, 244–249.
51. B. Esser, J. M. Schnorr and T. M. Swager, Selective detection of ethylene gas using carbon nanotube-based devices: utility in determination of fruit ripeness, *Angew. Chem.*, 2012, **51**, 5752–5756.
52. P. Chaudhary, F. Fatima and A. Kumar, Relevance of Nanomaterials in Food Packaging and its Advanced Future Prospects, *J. Inorg. Organomet. Polym.*, 2020, **30**, 5180–5192.
53. C. Sharma, R. Dhiman, N. Rokana and H. Panwar, Nanotechnology: An Untapped Resource for Food Packaging, *Front. Microbiol.*, 2017, DOI: 10.3389/fmicb.2017.01735.

54. A. Kumar, A. Choudhary, H. Kaur, S. Mehta and A. Husen, Metal-based nanoparticles, sensors, and their multifaceted application in food packaging, *J. Nanobiotechnol.*, 2021, **19**(1), 1–25.
55. T. Duncan, Applications of nanotechnology in food packaging and food safety: Barrier materials, antimicrobials and sensors, *J. Colloid Interface Sci.*, 2011, **363**(1), 1–24.
56. M. Shafiq, S. Anjum, C. Hano, I. Anjum and B. H. Abbasi, An overview of the applications of nanomaterials and nanodevices in the food industry, *Foods*, 2020, **9**(2), 148.
57. A. Sundramoorthy and S. Gunasekaran, Application of Graphene in quality assurance and safety of food, *TrAC, Trends Anal. Chem.*, 2020, **60**, 36–53.
58. N. Noor, Z. Ashraf and F. A. Masoodi, *Applications of Nanomaterials in Agriculture, Food Science, and Medicine*, 2021, p. 18.
59. J. Siddiqui, M. Taheri, A. U. Alam and M. J. Deen, Nanomaterials in Smart Packaging Applications: A Review, *Small*, 2022, **18**(1), 2101171.

CHAPTER 9

Nanotechnology to Detect the Microbial Toxins in Stored Food

AMRITA BHANJA, REETUPARNA NANDA AND MONALISA MISHRA*

Neural Developmental Biology Laboratory, Department of Life Science, National Institute of Technology, Rourkela 769008, Odisha, India
*Email: mishramo@nitrkl.ac.in

9.1 Introduction

With a rapidly increasing population, it is estimated that the global population may grow to 9 billion by 2050.[3] This will directly affect the food and agriculture sector. This sector not only feeds the population but also the main contributor to the economies of many developing countries.[4] This industry is currently estimated to be a multi-trillion-dollar industry worldwide. Despite all these factors, many problems affect the food industry. These problems include growing sufficient food on limited land and meeting the growing demands for other crops such as animal feed, biofuels and pharmaceuticals. Thus it is essential to protect food from microbial infestations and toxicity in order to minimise the losses in the food sector and efficiently use the limited amount of product for a huge population. Proper food packaging is also essential to maintain food safety and health.[5]

As food and agricultural products are consumed at several levels of processing it is important to assess the risks of toxins at each of these levels. Spoiling of food is a very common and frequent activity taking place during its processing, storing and transportation. Apart from insects and rodents,

Food Chemistry, Function and Analysis
Bio- and Nano-sensing Technologies for Food Processing and Packaging
Edited by Ashutosh Kumar Shukla

Published by the Royal Society of Chemistry, www.rsc.org

microorganisms, mostly bacteria and fungi, are the main cause of food spoilage, and the toxins secreted by them are the leading cause of food intoxication.[6] The major cause of foodborne outbreaks is a handful of Gram-positive bacteria including *Clostridium perfringens*, *Staphylococcus aureus*, *Clostridium botulinum* and *Bacillus cereus.* The toxins produced by these bacteria can be found at any stage of processing and they can be present in the food even after inactivation of the microorganism.[7] The toxins can cause acute to severe health issues in the consumer, hence they are one of the causes of food wastage during processing. Another cause of toxicity in stored food is the presence of mycotoxins. These are the secondary metabolites of fungi from the taxa *Ascomycota* which can be toxic in nature. Currently, more than 300 toxins have been identified from this phylum alone, which include zearalenone (ZEA), ochratoxins (OT), trichothecenes including deoxynivalenol (DON) and aflatoxins (AF). These toxins contaminate a great deal of food crops and animal feeds.[7,8] While species of *Fusarium* can be a threat to crops like barley, wheat and corn, species of *Penicillium* and *Aspergillus* grow mostly in stored food.[9,10] The fungal toxins have acute to chronic effects including teratogenic, nephrotoxic and immunotoxic effects.[11] While there are few available options to protect food from microbial infestations, nanotechnology is show promise for the future. Currently nanotechnology has been applied to various aspects of agriculture including aquaculture and food crops to increase the overall productivity.[4] Apart from these major causes there are some minor causes of toxicity in stored food such as algal toxins and toxins from released industrial effluents. After the rapid increase in industrialisation over the past few decades it has been noticed that land and water resources in India have become contaminated with toxic effluents from industry which may include heavy metals like mercury, cadmium, lead, arsenic and some organochlorine compounds like 1,1,1-trichloro-2, 2-bis(*p*-chlorophenyl)ethane (dichlorodiphenyl dichloroethene) (DDT). This can have dangerous effects on male and female reproductive systems and also results in neurodevelopmental toxicities in new-borns.[1,2] Even the application of pesticides on food crops can add toxicity to food materials as they contain organophosphates which have been scientifically proven to affect the developing brain. In recent years another source of toxicity has been identified in stored food, which is leaching from storage containers. Two major toxins that have been identified as causing problems for consumers are bisphenol A and phthalates. While the former is known to cause defects in male reproductive organs and the human central nervous system, the latter has been shown to induce defects in the male reproductive organs of lab animals.[11] As there is a wide range of food toxins, it calls for a wide variety of toxicity detection methods.

For instance, for the detection of mycotoxins, several detection methods such as high-performance liquid chromatography (HPLC), thin-layer chromatography (TLC), gas chromatography (GC), ultra-high performance liquid chromatography (UPLC) and enzyme-linked immunosorbent assay (ELISA) are used.[12,13] Some of the major challenges faced when using these quantification techniques are the detection of low-level mycotoxins and a large variety of chemical structures of the mycotoxins.[14–16] To tackle these challenges and to increase the efficiency of toxicity detection some scientists have used electrochemical affinity biosensors, especially for mycotoxin detection.[17] These sensors work on the basis of biochemical interactions which provide them with higher sensitivity. These methods are also affordable financially, which makes them a more popular choice for the detection of mycotoxins. The biosensors use carbon-based nanomaterials such as graphene and its derivatives (*e.g.* reduced graphene oxide) or multiwalled carbon nanotubes (MWCNTs).[18,19] They provide benefits such as higher sensitivity, higher specificity, higher operational stability, shorter analysis times, and simpler and integrated sample treatment.[20,21] Another toxin that possesses a serious threat and that is responsible for loss of food is the brevetoxins (BTXs). These are cyclic polyether by nature and are produced by dinoflagellates called *Karenia brevis.*[22] These toxins bind to voltage-gated sodium channels in the neurons, which causes disruption of function in the nerves commonly known as shellfish poisoning. To check this toxicity in seafood materials, a group of scientists have developed an electrochemical immunonanosensor that works on the theory of immobilised BTX–2eBSA conjugate on the gold nanoparticle-decorated amine-terminated poly(amidoamine) dendrimers (AuNPePAADs).[23]

9.2 Types of Methods Used to Detect Food Toxicity

Before the application of nanomaterials or nanoparticles for the analysis of food toxicity, there were some conventional methods that have been applied for the detection of toxins and foodborne pathogens present in food samples (Figure 9.1). Some of the most frequently used methods include PCR-based molecular methods, culture-based methods and immunoassays,[24] however these methods had many disadvantages and challenges such as time consuming, high cost, low specificity and sensitivity, labour intensive, low feasibility, *etc.*, eventually affecting the final outcome.[25] Some of the analytical methods used for the detection of food toxins are briefly discussed here. These techniques mostly include interactions and binding of specific antigens and antibodies which are achieved by assays such as immunoaffinity column assay (ICA), radioimmunoassay (RIA) and enzyme-linked immunosorbent assay (ELISA). As per the literature, RIA was used for the toxic analysis of peanuts for the detection of aflatoxin B1.[26] ELISA was used for the detection of *E. coli* and also *Shigella boydii* in several food samples including jelly, bread and milk.[27] Despite successful results, researchers had to deal with several challenges such as proper disposability of the radio-active isotopes,[28] requirement of pure antigen, time consuming, *etc.*, which questioned the whole process for the detection of food toxins. Other techniques, including the chromatographic techniques that include thin-layer chromatography (TLC) were basically used for determining the purity of several organic compounds, their minimum detection limit, and lowered utilisation of TLC.[29] According to Wacoo *et al.*,[30] HP-TLC, *i.e.*

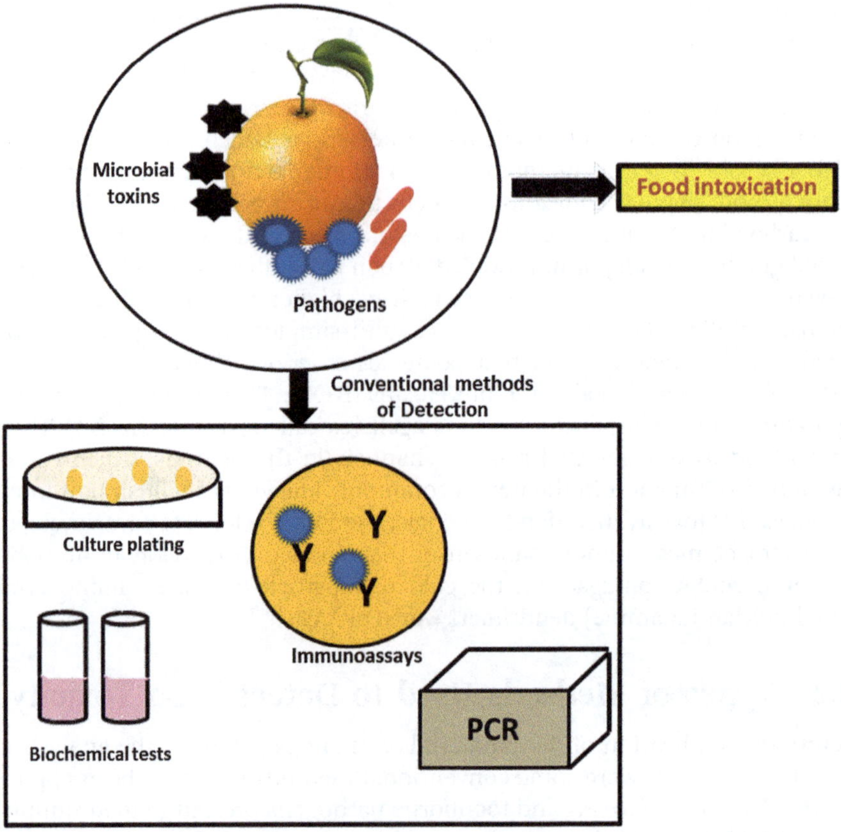

Figure 9.1 Schematic diagram showing the conventional methods used for the detection of pathogens.[25]

high-performance thin-layer chromatography, was also utilised for the process of toxin determination, but also had numerous limitations such as being expensive and requiring heavy instrumentation. Apart from these, the analysis process achieved by the liquid chromatography techniques generally include reversed-phase liquid chromatography, combined with pre-column derivatisation, reversed-phase liquid chromatography combined with electrochemical detection, and normal-phase liquid chromatography.[31] The liquid chromatography technique includes the involvement of several factors such as preparation of sample, cost and effect of the matrix. Also, high-performance liquid chromatography (HPLC) has been used widely in such detection processes and was found to enhance the sensitivity with good recovery, but on the other hand, it was also time consuming, with a large requirement of cleaning-up of the processes.[26] The disadvantages of these techniques led to the development of the utilisation of the biosensors modified with nanomaterials for the rapid and accurate detection of toxins and foodborne pathogens in food samples.

9.3 Types of Nanomaterials Used for Toxin Detection

Nanomaterials come with unique properties and are used in the form of capture probes that bind to the target analytes, resulting in detectable signals in biosensors. There are a variety of nanomaterials, of which some are discussed in this section.

9.3.1 Metal Nanoparticles

Gold and silver nanoparticles are among the metal nanoparticles and are present in a variety of sizes and shapes that have multifaceted optical properties controlled by localised surface plasmon resonance (LSPR). Due to the presence of LSPR, these nanoparticles exhibit an obvious colour that tends to change during the aggregation of particles or a change in the ambient dielectric constant. LSPR allows gold nanoparticles to increase the proximate fluorophore emission. These metal nanoparticles may also enhance the intensity of the Raman scattering by approximately up to 10^{14}-fold.[33]

9.3.2 Magnetic Nanoparticles

Iron-based nanoparticles are categorised under magnetic nanoparticles that includes the properties based on their shape, materials and size.[32,34] A particle size below 20 nm is considered to show a superparamagnetic effect that helps in preventing the aggregation of magnetic nanoparticles. Such kinds of nanoparticles are required for capturing and separating target analytes, while the detection limit is low.

9.3.3 Quantum Dots

These nanoparticles are considered to be semiconductive in nature, having a structure of a core–shell.[35] The quantum confinement effect based on the semiconducting band gap leads to fluorescence emission. This band gap can be controlled by altering the size and material arrangement of the quantum dots. As compared to organic fluorophores, quantum dots exhibit finer photostability, large quantum yield, with a broad absorbance spectrum. Quantum dots are used especially as fluorescent probes in biosensors.[36]

9.3.4 Silica Nanoparticles

These are nanoscale particles exhibiting high porosity that are used in biosensors as fluorescent probes as they have the ability to enfold a number of fluorophores in their matrix. This is the reason behind the high fluorescence intensity of silica nanoparticles as compared to single fluorophore molecules. The encapsulated fluorophores exhibit high photostability as they remain protected from the environmental conditions inside the silica matrix.[33]

9.4 Mode of Action of Nanomaterials Used for Food Sensing

The advancement of nanotechnology has revolutionised the field of biosensors. Nanobiosensors have been developed as second-generation biosensors with some great benefits such as the use of nanomaterials which has made these biosensors ultrasensitive. A group of scientists has illustrated the use of carbon nanotubes (CNTs) in biosensors. Nanowires are formed from CNTs which are used as biosensors.[37] Other types of nanoparticles that are used in biosensors are dendrimers, quantum dots, silica, gold and other metallic nanoparticles.[38] Nanomaterials can be used in biosensors either at the bioreceptor level or the transducer level.[39] Nanomaterials are better candidates for use at bioreceptor levels as they have a very high surface area as compared to others, which is a benefit in detecting toxins even at very low levels. Magnetic nanocomposites made from graphene oxide and other magnetic nanoparticles are used for the adsorption of aflatoxin M1.[40] This not only acts on biomolecules but also is very efficient in detecting microbes in stored food, which is more beneficial than older methods involving microbial cultures. The nanosensors are designed with the ability to bind specifically to a particular microorganism that is present in the food sample. It is often coupled with a nanosized light-scattering instrument for easy detection. Horner *et al.* effectively combined these techniques to detect *E. coli.*[41] In current situations we require real-time analysis of a number of microbes as well as the concurrent detection of multiple microbes in stored food, and these are efficiently achieved with the use of nanomaterials in biosensors.[42] Nanomaterials are also used at transducer levels where they act as an efficient quencher of the electron which ultimately helps in signal amplification.[43] For instance, Zhu *et al.* have shown the unique properties of gold nanoparticles in electrical signal amplification.[44]

9.5 Nanomaterials and Nanoparticles Used to Detect Common Food Toxins

The detection of food microbial toxins has been carried out using various methods which have been hindered by a number of challenges including low sensitivity, reduced specificity and longer detection time.[45] In order to overcome such challenges, nanomaterials combined with biosensors have been used. Some of the nanomaterial-based biosensors that are specifically designed to detect microbial toxins and pathogens include quantum dots, carbon nanotubes, carbon nanodrops and nanoparticles.[46]

The word nanomaterial has been well defined by the International Organisation for Standardisation. As per this organisation, it is a material ranging between 1–100 nm, which can be easily incorporated into biosensors resulting in several advantages including high sensitivity. In addition, nanomaterials are in high demand due to their compatible nature, stability

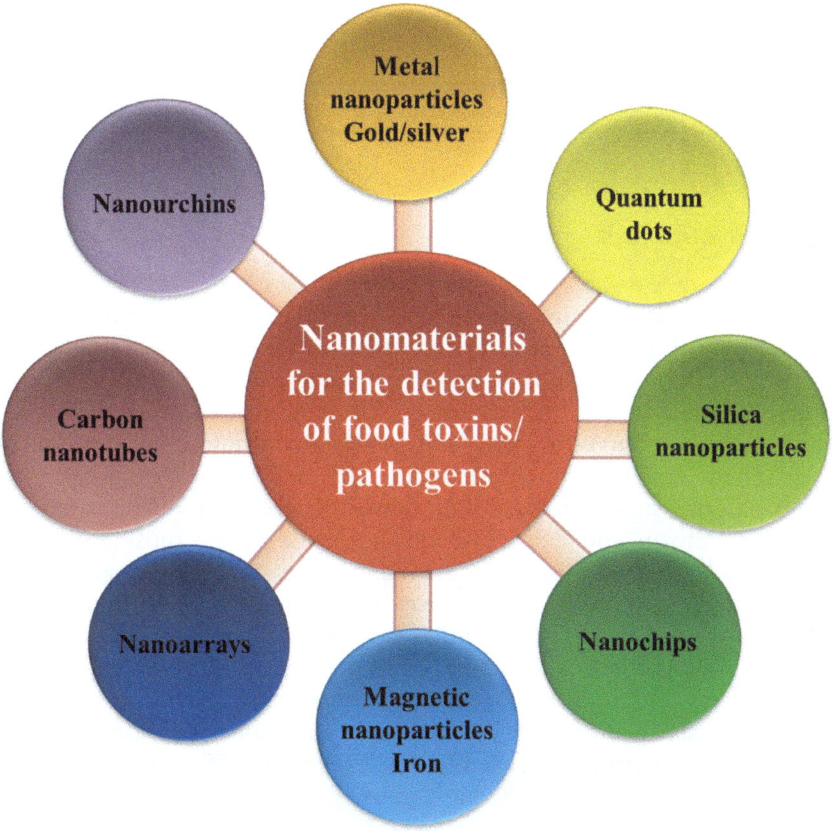

Figure 9.2 Types of nanomaterials used in the detection of microbial toxins and pathogens.[48]

and conductivity.[47] Nanomaterials have been combined with various electrochemical biosensors for successful detection of food microbial toxins (Figure 9.2). Before studying the nanomaterials used for microbial toxin detection, it is very important to have a brief knowledge of the type of microbial toxins that are responsible for increasing food toxicity.

9.6 Types of Microbial Toxins Affecting Food Quality

As per a literature survey, microbial toxins are the cause of 70% of foodborne disease outbreaks. These substances are produced by microbes such as bacteria, algae and fungi that may affect the health of the host by disrupting the plasma membrane. Their toxic effects can also be retained, affecting the cells even after their death.[49] The structure of the toxins may vary from small to macromolecules that generally include proteins and peptides. Therefore, the detection of microbial toxins has been an important field of research for several years. There are several ways to detect the toxins present in food matrices, as has been discussed earlier, however, apart from such methods, nanoparticle- or nanomaterial-based biosensors are in high demand due to their cost-effectiveness, ease of handling, reliability and high sensitivity. Some of the most common toxins that are found in the food matrices are described below.

9.6.1 Mycotoxins

Mycotoxins are the chemical substances produced by fungi. These are the most abundant and common form of toxins, and are mostly found in cereals, spices and dried fruits. These are the secondary metabolites produced by *Fusarium*, *Penicillium* and *Aspergillus*.[50] Generally environmental factors initiate the occurrence of these toxins. The food is contaminated by these toxins during the harvesting processes. These toxins affect cells by entering the lymphatic system through inhalation, skin or ingestion, causing cancer, immunosuppression, necrosis, inflammation, *etc.*[51] Aflatoxin, trichothecenes, orchratoxin and deoxynivalenol are some of the most well-known mycotoxins.[10] Aflatoxin is considered to be the most carcinogenic toxin amongst all the mycotoxins. They are mostly known for contaminating crop varieties and food types such as milk, rice, peanuts and corn, and may become the main source of contamination during the harvesting process.[52] Liver is the most highly targeted organ to be affected by such toxins[53] and in severe cases may lead to death when stored for a long period in the body. Therefore, such concerns have led to the development of various techniques that can detect toxin-producing pathogens.

9.6.2 Bacterial Toxins

The presence of bacteria in food is very common and is almost impossible to remove totally from food. Although not all bacteria are harmful, there are certain bacteria that produce toxins and are pathogenic. Some of the bacteria that are considered to be the primary sources of toxin production resulting in several foodborne illnesses such as pneumonia, diarrhoea, listeriosis, bacteraemia, *etc.* are *Escherichia coli*, *Clostridium perfringens*, *Listeria monocytogenes*, *Campylobacter* spp., *Staphylococcus* and *Salmonella*.[54] The toxins produced by bacteria are protein macromolecules that tend to be highly infectious. These bacterial toxins are of two types, enterotoxins and endotoxins. The pH and moisture content of the food play a very important role in maintaining its toxicity. As an example, *Clostridium botulinum* contaminates food with a pH higher than 4.5 with a high moisture content. *S. aureus* produces enterotoxins that are heat stable in nature, and the enterotoxins produced by *Staphylococcus* are the main cause of foodborne diseases.[26] Some other toxins responsible for foodborne disease outbreaks are the cholera toxin produced by *Vibrio cholera*, thermostable toxin, shiga toxin, beta toxin, alpha toxin and many more.[55]

9.6.3 Algal Toxins

These toxins mostly contaminate seafood. The toxins produced by algae are usually taken up by aquatic animals such as fish leading to the poisoning of the fish, which may further affect humans when consumed. As per a literature survey, the mortality rate due to the algal intoxication is around 1.5% per year.[56] Some temperature resistant algal toxins may last in the food for a longer period of time resulting in its spoilage. Algal toxins are categorised into two types of toxins: red tide toxins and cyanobacterial toxins. Neurotoxins and hepatotoxins are the two cyanobacterial toxins that contaminate fresh water by producing cyanobacterial blooms.

The above-mentioned are some of the common and well-established microbial toxins that frequently cause food spoilage resulting in foodborne disease outbreaks. In order to understand the compatibility or mechanism of the nanomaterials utilised in the detection of food toxins, it is necessary to have a brief idea of the types of toxins that are responsible for food intoxication. Therefore, in order reduce food intoxication, nanomaterials are being used to detect the toxins that affect the quality of the food.

9.7 Nanomaterial-based Biosensors Used for the Detection of Foodborne Pathogens

Food toxins can be easily monitored using a portable, sensitive device called a biosensor. Some of the biosensor-based methods have been approved by the European Food Safety Authority (EFSA) and some other organisations that use them for detecting the toxins present in food.[57] Nanomaterials are used specifically to enhance the sensitivity of biosensors. The nanomaterials that are most commonly used in such processes are gold, platinum and silver nanoparticles, nanosheets, nanochannels, quantum dots and nanowires.[58] Nanoparticles are used in detecting *Salmonella typhimurium* by carrying out the following measures, starting with fixing of the magnetic nanomaterials to the pathogenic cell, separating the excess iron oxide nanoparticles, and conjugating the pathogens with magnetic nanoparticles to the specific substrate in the magnetic field. The concentration of the pathogen can be determined by either determining the iron ion concentration in the solution or by detecting the status of the active polymer bound to the magnetic nanoparticles.[59] Immunosensors incorporated with zinc sulphide quantum dots are used to detect staphylococcal enterotoxin B. This was achieved by dissolving the immunocomplex in hydrochloric acid, which resulted in fluorescence of the zinc sulphide quantum dots.[60] Amongst all the *E. coli* strains, *E. coli* O157:H7 is considered to be the highest toxin-producing serotype that can damage the intestinal lining leading to several types of health conditions including stomach cramps, haemorrhagic colitis and anaemia.[61] Consuming raw products such as milk or undercooked meat are the main sources of infection. The detection of this specific serotype was carried out[62] using magnetic nanoparticles conjugated with streptavidin. It then was used to develop a DNA sensor. The detection process in milk was carried out using a single-use immunosensing strip. This strip contained double antibodies for carrying out enzyme-linked immunoassay and was modified by using gold nanoparticles. These gold nanoparticles increased the process sensitivity by 13.1-fold, resulting in the detection of 50 colony-forming units per strip in milk.[63] *E. coli* was successfully separated using Au nanoparticle-based iron oxide magnetic core–shells with a size of 12.5 nm.[64] The fast detection of *E. coli* was successfully carried out using carbon electrode reformed with platinum nanoparticles.[65] Due to its large surface area, the *E. coli* detection efficiency was greatly increased. The detection of *E. coli* and *Salmonella typhimurium*, the two best-known foodborne pathogens, was done in not more than 30 minutes. This process was carried out by functionalising amino-terminated gold nanodrops with antibodies.[66] In another experiment[67] magnetic nanoparticles were functionalised with D-mannose for the detection of *E. coli*. The detection was carried out by incubating the magnetic nanoparticles along with the concanavalin A labelled with fluorescein for 12 hours at 4 °C. Furthermore, it was again incubated with *E. coli* cells and separated with the help of a magnetic field. The listeriosis infection is caused by the *L. monocytogenes* bacterium which is known to be the most pathogenic foodborne bacterium and the third leading death-causing foodborne pathogen. The detection process of this pathogen was carried out by developing a high transition temperature base superconducting quantum interference device that would monitor the binding capacity of magnetic nanoparticles linked to the antibody and the bacterium. *L. monocytogenes* was combined with the antibody coated with the superparamagnetic nanoparticles. This process was followed by the application of the pulsed magnetic field leading to binding of the nanoparticles to *L. monocytogenes*.[68]

The magnetic nanoparticles, due to their unique characteristics such as cost effective, small size with large surface area and large stability, result in enhancement of the immunosensor sensitivity.[69] Silver and gold nanoparticles include magnificent plasmon absorption that increases the detection ability. Nanoparticles for the detection of toxins and pathogens have been extensively used, specifically bearing in mind the colorimetric sensing method. Appropriate utilisation of the colorimetric method was understood or carried out by modifying nanoparticles using relevant ligands. In order to carry out fast and systematic detection of microbial pathogens, poly *para*-phenyleneethynylene (PPE) combined with gold nanoparticles was utilised that resulted in PPE fluorescence due to the interaction of cationic Au nanoparticles and anionic PPE leading to the association of Au nanoparticles with the relevant pathogen further detaching the PPE and Au nanoparticles.[70] Hosseini and co-workers worked on the detection of Aflatoxin B1 using Au nanoparticles combined with the aptamers (specifically used as recognition ligand) and Au nanoparticles due to high absorption coefficient. The colour of Au nanoparticles changes from red to purple due to the aggregation of Au nanoparticles and desorption of Aflatoxin B1 binding aptamer from its surface.[71] Normally, magnetic nanoparticles are used in such pathogen detection processes as they carry the beneficial properties for efficient separation as compared to normal nanoparticles. These antibody-based magnetic nanoparticles were used to separate *Salmonella* from milk[72] by implementing an external magnetic field. Apart from these nanoparticles, quantum dots also are considered to be special nanoparticles and are extensively used in the biomedical field because of their tremendous optical characteristics.[73] As per a literature survey, there are several studies supporting the application of quantum dots in the form of fluorophores for detecting foodborne microbes. *Bacillus anthracis* pathogen was detected using a sandwiched nanoassembly made using a target DNA and two quantum dot nanoprobes.[74] Colorimetric assays, along with the nanoparticle probes, are very accurate and cost-effective and are also considered to be very effective nanotechnology-based techniques for the detection of toxins.[48]

9.8 Detection of Food Toxins Using Nanomaterial-based Sensors

There are several toxins responsible for food intoxication. These toxins are considered to be the protein complexes secreted by different microorganisms. Below various types of toxin responsible for foodborne illnesses and studies that have carried out the detection of such toxins are discussed.

9.8.1 Cholera Toxin

Vibrio cholera is the bacterium that produces cholera toxin, which causes diarrhoea during infection. In order to detect cholera toxin, Viswanathan *et al.*[75] developed a very practical method. This was a sandwich type assay that involved the binding of the toxin to the anti-cholera toxin antibody conjugated with ganglioside-functionalised liposome. The release of potassium ferrocyanide from the liposomes resulted in its binding to the electrode and it then was measured. Khan *et al.*[76] performed an experiment for the detection of cholera toxin using gold nanoparticles. Their experiment was based on dynamic light scattering and colorimetric assay. The resulting colour change due to the accumulation of antibody-conjugated gold nanoparticles in the presence of cholera toxin was visible. The detection of cholera toxin was also carried out by using microcantilever-based ganglioside nanodiscs.[77] This results in binding of the toxin to the nanodiscs through an active membrane protein. The optical interference is used for monitoring the nanomechanical cantilever response. This proved to be a promising method to detect cholera toxin.

9.8.2 Brevetoxins

Brevetoxins (BTXs) are made up of cyclic polyether compounds secreted naturally by *Karenia brevis* (a dinoflagellate species). These are marine neurotoxins. Egestion of this toxin leads to disorientation of the neurological processes resulting in neurological shellfish poisoning. Tang *et al.*[78] developed an electrochemical immunosensor that detected brevetoxin B in food samples. This immunosensor was developed by conjugating gold nanoparticle-based amine-terminated poly-amidoamine dendrimers with BTX-2 immobilised bovine serum albumin. This resulted in improved conductivity of the poly-amidoamine. A palladium nanostructure enriched with mesoporous carbon was developed by Lin *et al.*[79] for detecting brevetoxin B. This experiment was carried out by functionalising an electrode with brevetoxin B bovine serum albumin. This experiment became quite successful in the brevetoxin detection process, and was also applied to spiked marine food samples.

9.8.3 Shiga Toxin

Shiga toxin is produced by the most common foodborne microbes *E. coli* (O157:H7) and *Shigella dysenteriae*. Shiga toxin works by binding to the ribosomal subunit and preventing the production of protein. Quintela *et al.*[80] detected the shiga toxin producing *E. coli* using a gold nucleotide-based oligonucleotide. The single-stranded oligonucleotides modified with thiol were disabled on the gold nanoparticles followed by sandwiching the DNA samples collected from the food as well as pure culture with the gold nanoparticle oligo probes. This resulted in the formation of the complex due to hybridisation of the target DNA fragments and the gold nanoparticle probes. A change in the colour was observed signifying the presence of the concerned DNA in the test sample. Gold nanoparticles were used in the experiment conducted by Jyoti *et al.*[81] for the detection of *E. coli* nucleic acid. This was done by functionalising the 19 and 22 bp of single-stranded DNA with gold nanoparticles. The detection of shiga toxin was also successfully performed by Nagy *et al.*[82] using Gal-a1,4-Gal glycopolydiacetylene (nanoparticle) to develop a chromatic sensor. The resulting colour change from purple to brown indicated the detection of the shiga toxin. This method was found to be very specific and sensitive.

9.8.4 Staphylococcal Enterotoxin

Staphylococcal enterotoxin is one of the heat-resistant toxins secreted by the *Staphylococcus aureus* bacterium. The ingestion of this toxin may cause nausea, diarrhoea, vomiting and anorexia.[61] In addition, it may be the leading cause of rheumatoid arthritis, toxic shock syndrome and eczema.[63] The detection of staphylococcal enterotoxin using an enhanced chemiluminescence immunosensor modified with gold nanoparticles was developed by Yang *et al.*[83] This was followed by immobilising the anti-staphylococcal enterotoxin B antibodies to the gold nanoparticles with the help of physical adsorption, further immobilising the mixture (antibody–gold nanoparticle) on a polycarbonate surface. The toxin was detected through an ELISA sandwich assay by taking secondary antibody and enhanced chemiluminescence onto the polycarbonate surface. Another method using gold nanourchins for detecting staphylococcal enterotoxin B was carried out by Nodoushan *et al.*[84] They used an electrode modified with gold nanourchins and reduced grapheme oxide followed by attaching the DNA probe to the gold nanourchin surface. This resulted in the aptamer detachment from the surface of the electrode, indicating the presence of the staphylococcal enterotoxin B molecules. In another study, Mondal *et al.*[85] used staphylococcal enterotoxin B-binding aptamer and unmodified gold nanoparticles as the recognition element and colorimetric probes, respectively, for the detection of staphylococcal enterotoxin B. The detection of staphylococcal enterotoxin B was also performed by Wu *et al.*[86] using upconversion nanoparticles and platinum-coated gold nanorods in which the anti-staphylococcal enterotoxin B aptamer and the staphylococcal enterotoxin B aptamer were immobilised on platinum-coated gold nanorods and upconversion nanoparticles, respectively. The detection of staphylococcal enterotoxin A from milk was carried out by Becheva *et al.*[87] by immobilising antibodies on the magnetic nanoparticles modified with amino groups. In a study conducted by Belen *et al.*,[88] staphylococcal enterotoxin G was carried out utilising silica nanoparticles.

9.9 Conclusion

Nanotechnology has been a very challenging technology and it is used for several purposes. The use of nanotechnology in the food industry has been very specific and accurate, especially in the food analysis process. Starting from farming to marketing, the use of nanotechnology has brought about a great revolution.[89] In the food microbiology field, nanotechnology has played a crucial role in inhibiting the growth of pathogens by restricting their surface attachment and eliminating them. Nanomaterials are utilised for the production and inspection of food, by either modifying them or by conjugating them with some other materials, resulting in alteration of the physicochemical properties of the combined substances. In addition, they are also used for the detection of microbes, especially the pathogens and the toxins secreted by such pathogens responsible for the food spoilage.[90] The application of nanoparticles for such a purpose may affect or alter the elemental or molecular activity used for the determination of the pathogen and also may become the leading cause of the modification of the transport system of these organisms. Therefore, working with nanoparticles requires in-depth knowledge of the entry and mechanism of the nanoparticles within the cell, and the interaction mechanism of the cell and nanoparticles. This chapter also deals with the types and varieties of toxins and pathogens frequently affecting the quality of food. Basic information regarding the detection of toxins using nanomaterials is provided in this chapter.

References

1. U. R. Acharya, S. Acharya and M. Mishra, Lead acetate induced cytotoxicity in male germinal cells of Swiss mice, *Ind. Health*, 2003, **41**(3), 291–294.
2. U. R. Acharya, M. Mishra and I. Mishra, Status of antioxidant defense system in chromium-induced Swiss mice tissues, *Environ. Toxicol. Pharmacol.*, 2004, **17**(3), 117–123.
3. A. Ghasemzadeh, Global issues of food production, *Agrotechnol*, 2012, **1**(2), 1–2.
4. B. S. Sekhon, Nanotechnology in agri-food production: an overview, *Nanotechnol., Sci. Appl.*, 2014, **7**, 31.
5. S. Mihindukulasuriya and L.-T. Lim, Nanotechnology development in food packaging: A review, *Trends Food Sci. Technol.*, 2014, **40**(2), 149–167.
6. S. T. Hammond, *et al.*, Food spoilage, storage, and transport: Implications for a sustainable future, *BioScience*, 2015, **65**(8), 758–768.
7. A. Rajkovic, Microbial toxins and low level of foodborne exposure, *Trends Food Sci. Technol.*, 2014, **38**(2), 149–157.
8. V. Pereira, J. Fernandes and S. Cunha, Mycotoxins in cereals and related foodstuffs: A review on occurrence and recent methods of analysis, *Trends Food Sci. Technol.*, 2014, **36**(2), 96–136.
9. T. Tanaka, *et al.*, Worldwide contamination of cereals by the Fusarium mycotoxins nivalenol, deoxynivalenol, and zearalenone. 1. Survey of 19 countries, *J. Agric. Food Chem.*, 1988, **36**(5), 979–983.
10. J. W. Bennett and M. Klich, Mycotoxins, *Clin. Microbiol. Rev.*, 2003, **16**(3), 497–516.
11. A. Borchers, *et al.*, Food safety, *Clin. Rev. Allergy Immunol.*, 2010, **39**(2), 95–141.
12. R. Krska, *et al.*, Mycotoxin analysis: An update, *Food Addit. Contam.*, 2008, **25**(2), 152–163.
13. N. W. Turner, *et al.*, Analytical methods for determination of mycotoxins: An update (2009–2014), *Anal. Chim. Acta*, 2015, **901**, 12–33.
14. H. Yazdanpanah, Mycotoxins: analytical challenges, *Iran. J. Pharm. Res.*, 2011, **10**(4), 653.
15. G. S. Shephard, Current status of mycotoxin analysis: a critical review, *J. AOAC Int.*, 2016, **99**(4), 842–848.
16. J. Stroka and C. Maragos, Challenges in the analysis of multiple mycotoxins, *World Mycotoxin J.*, 2016, **9**(5), 847–861.
17. J. C. Vidal, *et al.*, Electrochemical affinity biosensors for detection of mycotoxins: A review, *Biosens. Bioelectron.*, 2013, **49**, 146–158.
18. C. Singh, *et al.*, Carboxylated multiwalled carbon nanotubes based biosensor for aflatoxin detection, *Sens. Actuators, B*, 2013, **185**, 258–264.
19. S. Srivastava, *et al.*, Electrophoretically deposited reduced graphene oxide platform for food toxin detection, *Nanoscale*, 2013, **5**(7), 3043–3051.
20. J. Wang, Carbon-nanotube based electrochemical biosensors: A review, *Electroanalysis*, 2005, **17**(1), 7–14.
21. T. Kuila, *et al.*, Recent advances in graphene-based biosensors, *Biosens. Bioelectron.*, 2011, **26**(12), 4637–4648.

22. S. M. Plakas and R. W. Dickey, Advances in monitoring and toxicity assessment of brevetoxins in molluscan shellfish, *Toxicon*, 2010, **56**(2), 137–149.
23. Z. Wang, *et al.*, LC/MS analysis of brevetoxin metabolites in the Eastern oyster (Crassostrea virginica), *Toxicon*, 2004, **43**(4), 455–465.
24. N. Bhardwaj, *et al.*, Fluorescent nanobiosensors for the targeted detection of foodborne bacteria, *TrAC, Trends Anal. Chem.*, 2017, **97**, 120–135.
25. V. Velusamy, *et al.*, An overview of foodborne pathogen detection: In the perspective of biosensors, *Biotechnol. Adv.*, 2010, **28**(2), 232–254.
26. R. Gupta, *et al.*, Advances in nanomaterial-based electrochemical biosensors for the detection of microbial toxins, pathogenic bacteria in food matrices, *J. Hazard. Mater.*, 2021, **401**, 123379.
27. C. Song, *et al.*, Development of a lateral flow colloidal gold immunoassay strip for the simultaneous detection of Shigella boydii and Escherichia coli O157: H7 in bread, milk and jelly samples, *Food Control*, 2016, **59**, 345–351.
28. R. Twyman, *Immunoassays, Applications/Clinical*, 2005.
29. L. Zhang, *et al.*, Contribution of blocking positions on the curing behaviors, networks and thermal properties of aromatic diamine-based benzoxazines, *Thermochim. Acta*, 2018, **668**, 65–72.
30. A. P. Wacoo, *et al.*, Methods for detection of aflatoxins in agricultural food crops, *J. Appl. Chem.*, 2014, **1–15**, 706291.
31. A. Rahmani, S. Jinap and F. Soleimany, Qualitative and quantitative analysis of mycotoxins, *Compr. Rev. Food Sci. Food Saf.*, 2009, **8**(3), 202–251.
32. A. Malachová, *et al.*, Advanced LC–MS-based methods to study the co-occurrence and metabolization of multiple mycotoxins in cereals and cereal-based food, *Anal. Bioanal. Chem.*, 2018, **410**(3), 801–825.
33. L. Sutarlie, S. Y. Ow and X. Su, Nanomaterials-based biosensors for detection of microorganisms and microbial toxins, *Biotechnol. J.*, 2017, **12**(4), 1–25.
34. H. Lee, *et al.*, Recent developments in magnetic diagnostic systems, *Chem. Rev.*, 2015, **115**(19), 10690–10724.
35. D. Vasudevan, *et al.*, Core–shell quantum dots: Properties and applications, *J. Alloys Compd.*, 2015, **636**, 395–404.
36. M. Mishra, Latest Trends in Bioimaging Using Quantum Dots, in *Smart Nanodevices for Point-of-Care Applications*, CRC Press, 2022, pp. 135–142.
37. E. Katz and I. Willner, Biomoleculeifunctionalized carbon nanotubes: applications in nanobioelectronics, *Chem. Phys. Chem.*, 2004, **5**(8), 1084–1104.
38. S. Otles and B. Yalcin, Nano-biosensors as new tool for detection of food quality and safety, *LogForum*, 2010, **6**(4), 7.
39. M. Rai, P. S. Jogee and A. P. Ingle, Emerging nanotechnology for detection of mycotoxins in food and feed, *Int. J. Food Sci. Nutr.*, 2015, **66**(4), 363–370.
40. N. Gan, *et al.*, An ultrasensitive electrochemiluminescent immunoassay for Aflatoxin M1 in milk, based on extraction by magnetic graphene and detection by antibody-labeled CdTe quantumn dots-carbon nanotubes nanocomposite, *Toxins*, 2013, **5**(5), 865–883.

41. S. R. Horner, *et al.*, A proteomic biosensor for enteropathogenic E. coli, *Biosens. Bioelectron.*, 2006, **21**(8), 1659–1663.
42. H. Yang, H. Li and X. Jiang, Detection of foodborne pathogens using bioconjugated nanomaterials, *Microfluid. Nanofluid.*, 2008, **5**(5), 571–583.
43. M. R. Felício, *et al.*, Peptides with dual antimicrobial and anticancer activities, *Front. Chem.*, 2017, **5**, 5.
44. K. Zhu, *et al.*, Recent developments in antibody-based assays for the detection of bacterial toxins, *Toxins*, 2014, **6**(4), 1325–1348.
45. O. Lazcka, F. Javier Del Campo and F. Xavier Muñoz, Pathogen detection: A perspective of traditional methods and biosensors, *Biosens. Bioelectron.*, 2007, **22**(7), 1205–1217.
46. I.-H. Cho, *et al.*, Nano/micro and spectroscopic approaches to food pathogen detection, *Annu. Rev. Anal. Chem.*, 2014, **7**, 65–88.
47. L. Reverté, B. Prieto-Simón and M. Campàs, New advances in electrochemical biosensors for the detection of toxins: Nanomaterials, magnetic beads and microfluidics systems. A review, *Anal. Chim. Acta*, 2016, **908**, 8–21.
48. C. Kaittanis, S. Santra and J. M. Perez, Emerging nanotechnology-based strategies for the identification of microbial pathogenesis, *Adv. Drug Delivery Rev.*, 2010, **62**(4–5), 408–423.
49. M. Pohanka, Current Trends in the Biosensors for Biological Warfare Agents Assay, *Materials*, 2019, **12**(14), 2303.
50. S. V. Avery, *et al.*, The fungal threat to global food security, *Fungal Biol.*, 2019, **123**(8), 555–557.
51. E. Wielogorska, *et al.*, Occurrence and human-health impacts of mycotoxins in Somalia, *J. Agric. Food Chem.*, 2019, **67**(7), 2052–2060.
52. C. Yan, *et al.*, Recent advances in aflatoxins detection based on nanomaterials, *Nanomaterials*, 2020, **10**(9), 1626.
53. J. Misihairabgwi, *et al.*, Mycotoxin contamination of foods in Southern Africa: A 10-year review (2007–2016), *Crit. Rev. Food Sci. Nutr.*, 2019, **59**(1), 43–58.
54. L. Chen and W. Alali, Recent discoveries in human serious foodborne pathogenic bacteria: resurgence, pathogenesis, and control strategies, *Front. Microbiol.*, 2018, **9**, 2412.
55. C. Hernández-Cortez, *et al.*, Food poisoning caused by bacteria (food toxins), *Poisoning: From Specific Toxic Agents to Novel Rapid and Simplified Techniques for Analysis*, 2017, p. 33.
56. X. Wu *et al.*, Application of novel nanomaterials for chemo-and biosensing of algal toxins in shellfish and water, *Novel Nanomaterials for Biomedical, Environmental and Energy Applications*, Elsevier, 2019, pp. 353–414.
57. B. D. Malhotra, S. Srivastava and S. Augustine, Biosensors for food toxin detection: Carbon nanotubes and graphene, *MRS Online Proc. Libr.*, 2015, 1725.
58. K. Brainina, *et al.*, 6. Nanomaterials: Electrochemical Properties and Application in Sensors, *Nanoanalytics*, 2018, 165–222.
59. K. Z. Brainina, A. N. Kozitsina and Y. A. Glazyrina, Hybrid electrochemical/magnetic assay for Salmonella typhimurium detection, *IEEE Sens. J.*, 2010, **10**(11), 1699–1704.

60. A. Sharma, *et al.*, Relative efficiency of zinc sulfide (ZnS) quantum dots (QDs) based electrochemical and fluorescence immunoassay for the detection of Staphylococcal enterotoxin B (SEB), *Biotechnol. Rep.*, 2015, **6**, 129–136.
61. S. K. Sonawane, *et al.*, Use of nanomaterials in the detection of food contaminants, *Eur. J. Food Res. Rev.*, 2014, 301–317.
62. X. Mao, *et al.*, A nanoparticle amplification based quartz crystal microbalance DNA sensor for detection of Escherichia coli O157: H7, *Biosens. Bioelectron.*, 2006, **21**(7), 1178–1185.
63. B. S. Inbaraj and B. Chen, Nanomaterial-based sensors for detection of foodborne bacterial pathogens and toxins as well as pork adulteration in meat products, *J. Food Drug Anal.*, 2016, **24**(1), 15–28.
64. U. Tamer, *et al.*, Synthesis of magnetic core–shell Fe_3O_4–Au nanoparticle for biomolecule immobilization and detection, *J. Nanopart. Res.*, 2010, **12**(4), 1187–1196.
65. Y. Cheng, *et al.*, Combining biofunctional magnetic nanoparticles and ATP bioluminescence for rapid detection of Escherichia coli, *Talanta*, 2009, 77(4), 1332–1336.
66. C. Wang and J. Irudayaraj, Gold nanorod probes for the detection of multiple pathogens, *Small*, 2008, **4**(12), 2204–2208.
67. K. El-Boubbou, C. Gruden and X. Huang, Magnetic glyco-nanoparticles: a unique tool for rapid pathogen detection, decontamination, and strain differentiation, *J. Am. Chem. Soc.*, 2007, **129**(44), 13392–13393.
68. H. Grossman, *et al.*, Detection of bacteria in suspension by using a superconducting quantum interference device, *Proc. Natl. Acad. Sci. U. S. A.*, 2004, **101**(1), 129–134.
69. S. Pal and E. C. Alocilja, Electrically active magnetic nanoparticles as novel concentrator and electrochemical redox transducer in Bacillus anthracis DNA detection, *Biosens. Bioelectron.*, 2010, **26**(4), 1624–1630.
70. R. L. Phillips, *et al.*, Rapid and efficient identification of bacteria using gold-nanoparticle–poly (*para*-phenyleneethynylene) constructs, *Angew. Chem., Int. Ed.*, 2008, **47**(14), 2590–2594.
71. M. Hosseini, *et al.*, Aptamer-based colorimetric and chemiluminescence detection of aflatoxin B1 in foods samples, *Acta Chim. Slov.*, 2015, **62**(3), 721–728.
72. H. HuiáShin and H. JoonáCha, A facile and sensitive detection of pathogenic bacteria using magnetic nanoparticles and optical nanocrystal probes, *Analyst*, 2012, **137**(16), 3609–3612.
73. M. Lv, *et al.*, Engineering nanomaterials-based biosensors for food safety detection, *Biosens. Bioelectron.*, 2018, **106**, 122–128.
74. Y.-P. Ho, *et al.*, Multiplexed hybridization detection with multicolor colocalization of quantum dot nanoprobes, *Nano Lett.*, 2005, **5**(9), 1693–1697.
75. S. Viswanathan, *et al.*, Electrochemical immunosensor for cholera toxin using liposomes and poly (3, 4-ethylenedioxythiophene)-coated carbon nanotubes, *Anal. Chem.*, 2006, **78**(4), 1115–1121.

76. S. A. Khan, *et al.*, Rapid and sensitive detection of cholera toxin using gold nanoparticle-based simple colorimetric and dynamic light scattering assay, *Anal. Chim. Acta*, 2015, **892**, 167–174.
77. S.-H. Tark, *et al.*, Nanomechanical detection of cholera toxin using microcantilevers functionalized with ganglioside nanodiscs, *Nanotechnology*, 2010, **21**(43), 435502.
78. D. Tang, *et al.*, Gold nanoparticles-decorated amine-terminated poly (amidoamine) dendrimer for sensitive electrochemical immunoassay of brevetoxins in food samples, *Biosens. Bioelectron.*, 2011, **26**(5), 2090–2096.
79. Y. Lin, *et al.*, Mesoporous carbon-enriched palladium nanostructures with redox activity for enzyme-free electrochemical immunoassay of brevetoxin B, *Anal. Chim. Acta*, 2015, **887**, 67–74.
80. I. A. Quintela, *et al.*, Simultaneous direct detection of Shiga-toxin producing Escherichia coli (STEC) strains by optical biosensing with oligonucleotide-functionalized gold nanoparticles, *Nanoscale*, 2015, **7**(6), 2417–2426.
81. A. Jyoti, *et al.*, Colorimetric detection of nucleic acid signature of shiga toxin producing Escherichia coli using gold nanoparticles, *J. Nanosci. Nanotechnol.*, 2010, **10**(7), 4154–4158.
82. J. O. Nagy, *et al.*, Glycopolydiacetylene nanoparticles as a chromatic biosensor to detect Shiga-like toxin producing Escherichia coli O157: H7, *Bioorg. Med. Chem. Lett.*, 2008, **18**(2), 700–703.
83. M. Yang, *et al.*, Gold nanoparticle-based enhanced chemiluminescence immunosensor for detection of Staphylococcal Enterotoxin B (SEB) in food, *Int. J. Food Microbiol.*, 2009, **133**(3), 265–271.
84. S. M. Nodoushan, *et al.*, An electrochemical aptasensor for staphylococcal enterotoxin B detection based on reduced graphene oxide and gold nano-urchins, *Biosens. Bioelectron.*, 2019, **127**, 221–228.
85. B. Mondal, *et al.*, Highly sensitive colorimetric biosensor for staphylococcal enterotoxin B by a label-free aptamer and gold nanoparticles, *Front. Microbiol.*, 2018, **9**, 179.
86. Z. Wu, D. He and B. Cui, A fluorometric assay for staphylococcal enterotoxin B by making use of platinum coated gold nanorods and of upconversion nanoparticles, *Microchim. Acta*, 2018, **185**(11), 1–8.
87. Z. Becheva, *et al.*, Rapid immunofluorescence assay for staphylococcal enterotoxin A using magnetic nanoparticles, *Int. J. Food Sci. Technol.*, 2019, **54**(3), 916–922.
88. S. M. Belen, *et al.*, Optimized surface plasmon resonance immunoassay for staphylococcal enterotoxin G detection using silica nanoparticles, *Biochem. Biophys. Res. Commun.*, 2021, **558**, 168–174.
89. N. Dasgupta, *et al.*, Nanotechnology in agro-food: from field to plate, *Food Res. Int.*, 2015, **69**, 381–400.
90. I. Bata-Vidács, *et al.*, Nanotechnology and microbial food safety, *Microbial pathogens and strategies for combating them: science, technology and education*, Formatex, Zubaran, 2013, pp. 155–159.

CHAPTER 10

Precautionary Measures for Developing Nanosensors for the Food Industry

SHARDA SUNDARAM SANJAY

Department of Chemistry, Ewing Christian College, Allahabad-211003, India
Email: sharda100@redifmail.com

10.1 Introduction

This is the era of nanotechnology. The majority of studies have been done on nanomaterials that have properties that differ from the same chemical composition albeit with relatively larger particles. Their potential in the food nanotechnology field is now being reviewed. Commercial foods and food supplements with incorporated nanoparticles are becoming available all over the world. In the global competition to use nanotechnologies in food production, processing, and packaging, huge amounts of money are being spent.

Natural nanostructures in foods are being studied primarily to enhance the food's functional nature. The potential of nanotech food seems to be endless.

Food companies have been intrigued by nanotechnology's potential implications, such as detecting infections in packaging or creating better flavours and colourings.

Presently, there are only a few foodstuffs that use nanotechnology in any form. Nanotechnology has the potential to transform the food system and

Food Chemistry, Function and Analysis
Bio- and Nano-sensing Technologies for Food Processing and Packaging
Edited by Ashutosh Kumar Shukla

Published by the Royal Society of Chemistry, www.rsc.org

have a significant impact on food science, since it could lead to improvements in food texture, flavour, process ability, and shelf-life sustainability. If the challenges it raises are overcome, it has the potential to improve food safety, quality, and shelf life.[1]

Even with such extraordinary properties, nanoparticles are vastly different entities that offer unexpected and unusual threats to human health and the environment, needing appropriate safety measures.

Bionanotechnology explores the connections between biology and nanotechnology by attempting to uncover the fundamental principles behind biological processes as well as the development of instruments for extremely precise control of matter at the nanoscale level. Bionanotechnology is a truly interdisciplinary field that combines the physical and life sciences, as well as engineering. Biosensors and lab-on-a-chip devices are common bionanotechnological outputs.[2]

As foodstuffs have such a high nutritional value, that makes them susceptible to microbial infections. Microorganisms contaminate food and produce toxins as a form of self-defence, causing an array of ailments in humans. Their health is also adversely affected as a result of food tampering. Toxins and adulterants can cause illnesses from vomiting to diarrhoea to paralysis and sometimes even death. Food safety is thus a major global concern, because food is a key and essential need of humans.

As an ultimate requisite, the development of rapid and specific methods to identify dangerous chemicals, microorganisms, and associated toxins is essential.[3–5]

Nanoparticles are progressively being used to develop monitoring techniques for detecting contamination, adulteration, and the freshness of food products due to their exceptional attributes. Food products that are properly packaged are protected from moisture, infection, and spoiling. Conventional inert packaging serves as an air, dust, and moisture barrier. Passive packaging techniques, on the other hand, fall short of meeting the growing concerns about food safety and biodefence.[6]

Intelligent packaging systems, on the other hand, can detect the quality of food articles and safeguard the shelf life and brand identity of packaged food. Intelligent packaging makes use of barcodes, time–temperature indications, gas indicators, and biosensors, *etc.* Conventional barcode-based protection tags can be effortlessly manipulated, but nanoparticle-based invisible and sophisticated tags are very difficult to crack and could be very beneficial in the future for verifying the authenticity of food products.[7,8] Thus, sensing is a crucial component of an intelligent packaging system.[9]

Microbial contamination, hazardous chemicals, and pesticides may all be detected quickly, sensitively, and reliably using nanosensors and nanopackaging materials. A recent breakthrough in nanotechnology has spawned a slew of nano-based scientific and industrial fields, with the market for nanomaterial-based goods growing steadily.

Despite its many benefits, the growing application of nanotechnology in food technology has prompted concerns about public safety, as well as ethical policies and regulatory concerns. Nanomaterials may have significantly different physicochemical and biological properties than their conventional form, and these unidentified features may produce unusual damage when they interact directly with humans through oral consumption. Despite the fact that bioactive component nanoencapsulation technology has been extensively investigated in the food sector, the fate of nanocarriers in the gastrointestinal tract varies greatly depending on their susceptibility to hydrolysis by digestive enzymes and the environment of the gastrointestinal tract.[10] Free nanocarriers, at the same time, frequently bypass intestinal/cellular boundaries, potentially increasing the bioaccumulation of foreign compounds in human blood, cells, and tissues. High use of organic solvents and surfactants in nanocarrier synthesis could provide a risk due to evaporation and it can also result in unpredicted residual solvents lingering in the final product, posing a potential hazard if their quantity is uncertain. The World Health Organization (WHO), the Food and Drug Administration (FDA), and the European Food Safety Authority (EFSA) have all recognized solvents and surfactants as hazardous, with safe usage levels established.[11]

Actually, there is a dearth of proper information on the potential safety issues associated with nanotechnology. The safety of nanoparticles and their associated dangers have yet to be studied in a proper way, necessitating deeper risk assessment. Nanomaterials' direct and indirect implications on human health must be investigated, as well as the biological fate of nanomaterials after digestion, their activity inside the gastrointestinal tract, and their potential interactions with biological systems. Furthermore, developing regulatory restrictions to safeguard the public from possible adverse impacts of nanotechnology is critical.[12] The toxicological effects of nanoparticles have been studied in a variety of ways. Many negative consequences have been observed, particularly in animals, despite the fact that the different toxicological aspects and the diversity of nanomaterials examined are only just getting started. The major cause of concern is insoluble or low-solubility nanoparticles. The toxicity of nanoparticles appears to be linked to their surface, which is a unique property.

This is a significant deviation from the usual, in which toxicity is mainly associated with a product's bulk.

Nanoparticles are so minuscule that even modest amounts (measured in mass) can have significant hazardous effects due to their high surface area.

Because of the small number of studies, the short exposure period, the different compositions of the nanoparticles tested (diameter, length, and agglomeration), and the frequently unexpected exposure route in the workplace, among other factors, toxicological data specific to nanoparticles remain insufficient.

To estimate the danger associated with inhalation and percutaneous exposure to humans, more research is needed on absorption, transfer to other tissues or organs, bio-persistence, carcinogenicity, and so on.

Realizing that the diversity, amount, quality, commercial production, and use of accessible nanomaterials will grow as new uses are created, the inadequacies of the data now available are particularly significant. Currently, there are no scientific procedures in practice to analyse the toxicity of each of these newer products in a timely and cost-effective manner.

Nanotechnology has a great deal of potential for providing benefits for food products, opening new opportunities for food innovations at a breakneck pace, but it also raises questions about safety and health. Therefore, as a precautionary measure, we must first consider all of the pros and cons of the newer technology while developing nanosensors for the food industry.

To conduct a thorough investigation into precautionary measures for developing nanosensors for the food industry, we must proceed step by step, beginning with a brief overview of sensors, nanosensors, and their applications in the food industry, as well as the results of their exposure to human health and then preventive measures. We start with sensors.

10.2 What is a Sensor?

Sensors are devices that perform an 'input' function by 'sensing' a physical change in some characteristic of a substance (analyte) in response to the changes in its surroundings such as temperature, pH, humidity, water flow, intensity of light, and so on. Transducers capture this change and convert it into a measurable or observable signal, thereby converting one type of energy into another. A transducer, as an electrical transducer, translates sound waves into electrical signals for amplification (a process) by an amplifier, and a loudspeaker (output device) converts these electrical signals back into sound waves. Figure 10.1 depicts a basic sensor configuration.

10.2.1 Biosensors

In a similar way, a biosensor is a type of analytical sensing device that is comprised of a biological ingredient, such as a cell, enzyme, or antibody, that acts as a bioreceptor, connected to a tiny transducer, the detecting part, *i.e.* a device that changes one type of signal into another so that it can be measured by another system, like a responsive displaying unit or a signal conditioning electrical circuit. Any biological element, that can be a tissue,

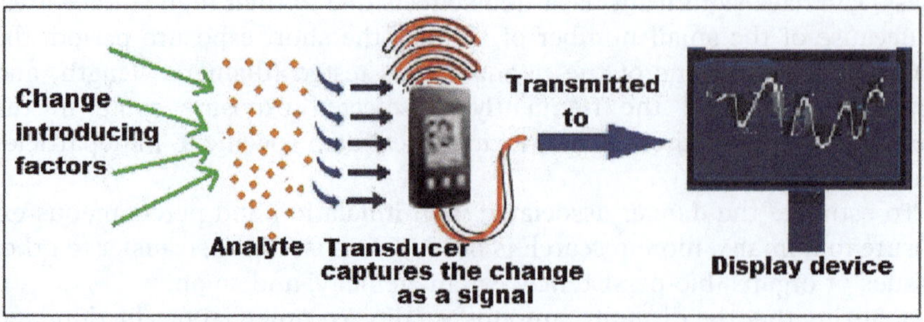

Figure 10.1 A simple representation of the working of a sensor.

microorganisms, organelles, enzymes, antibodies or nucleic acids, *etc.* which identifies the test analyte, may serve as a bioreceptor. On the basis of the type of the biomolecules, bioreceptors can be divided into six groups:

1. Protein based (mostly antibody/antigen based);
2. Enzyme based;
3. Phage based;
4. Nucleic acid DNA based;
5. Using complete cells or cell organelles as a basis for cellular interactions;
6. Making use of biomimetic synthetic bioreceptors.

The transducer, as mentioned above, is a physicochemical detector component that transforms the identified characteristics into a quantifiable signal.[13,14] The biochemical uniqueness of the biologically active substance determines the biosensor's function. The biological material chosen will be determined by a variety of parameters including selectivity, stowage, functional, and ambient stability.[13,15] A biosensor tenders its usage for sensing a number of applications, namely biomedical, industrial, in blood glucose sensing, and military applications. Some of its applications are included in Figure 10.2.

10.2.2 Functioning of Biosensors

Generally, a specific enzyme or selected biological material is deactivated by any common means and kept in close proximity to the transducer. The analyte connects to the biological object to recognize a clear analyte signal, which may be a colour change or evolution of heat, electron, gas, or hydrogen ion discharge. This, in turn, is recorded by the recording device available.

10.2.3 Kinds of Biosensors

Biosensors are an important component of sensing devices since they are directly responsible for the assay's bioanalytical performance as they convert or transduce a biological reaction into a detectable signal. Quantifiable signals, such as optical, electrochemical, piezoelectric, or thermal, have high sensitivity, precision, and low detection limits. They also offer a lot of potential in real-time analysis.

Biosensors are classified based on their quantitative approaches (Figure 10.3), as follows.

10.2.3.1 Thermal Biosensors

This type of biosensor takes advantage of one of the most basic characteristics of biological processes, *i.e.* heat absorption or generation, which

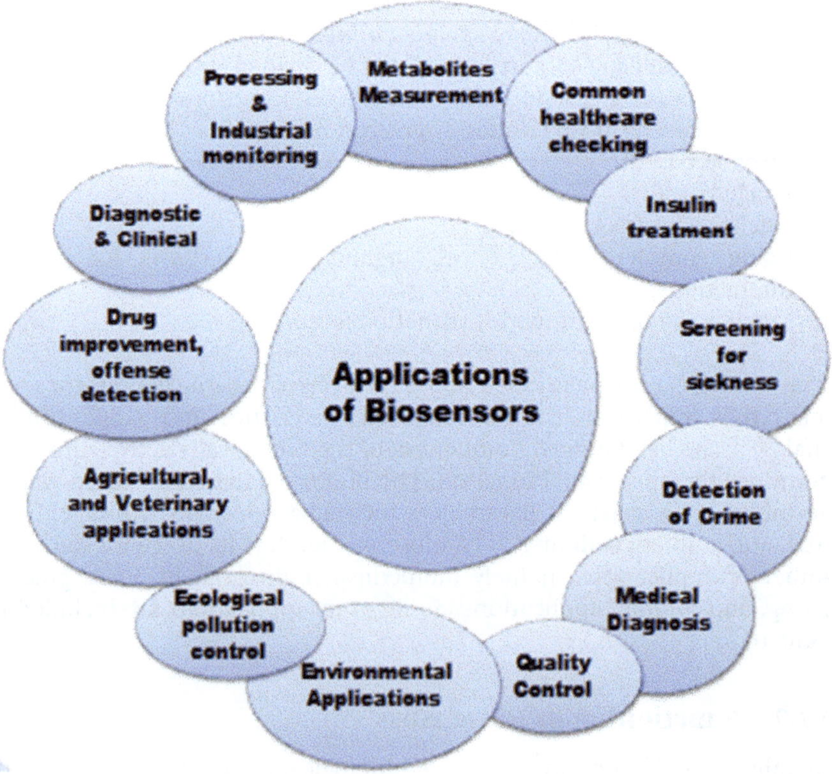

Figure 10.2 Biosensor applications.

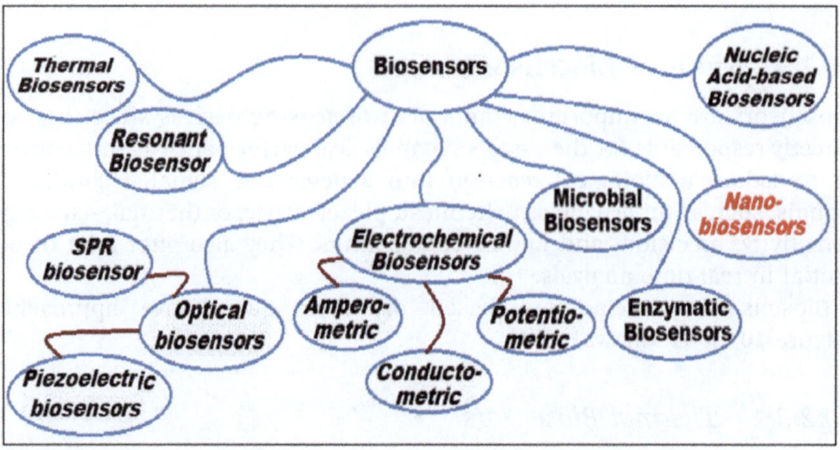

Figure 10.3 Schematic representation of various types of biosensors.

changes the temperature of the medium in which the response takes place. This is done by mounting enzyme molecules together with temperature sensors. The heat response of the enzyme is monitored and validated against the concentration of the analyte whenever the analyte encounters the enzyme. For the detection of pesticides and pathogenic microbes, this type of biosensor is employed.[16]

10.2.3.2 Resonant Biosensors

In this form of biosensor, an acoustic wave transducer is integrated with only an antibody, a biological entity. The mass of the biosensor membrane varies when such an analyte antigen molecule is coupled to it. The transducer's resonance frequency is changed as a result of the change in mass. Thereafter, the variation in frequency is recorded.[17]

10.2.3.3 Piezoelectric Biosensors

The capacity of certain materials to produce an electric response to an applied physical loading is known as the 'piezoelectric effect'. The piezoelectric (PZ) biosensor has a real-time output, is easy to use, and is inexpensive. The basic concept is to coat the PZ sensor's surface with a specifically binding agents, such as antibacterial antibodies, and then place it in a solution that contains bacteria. The bacteria connect to the antibodies, increasing the crystal's bulk whilst decreasing proportionally the resonance frequency of oscillation.[17]

10.2.3.4 Optical Biosensors

Optical scattering or electro-chemiluminescence is used to create such biosensors. This sort of biosensor measures the output transduced signal, which really is light. Optical biosensors are based on optical signal transduction and include ultraviolet, visible, and infrared spectrophotometry in transmission and reflection modes. Lambert–Beer's law readily resolves the correlation between incident light intensity and transmitted radiation. Optical phenomena such as absorption, refractive indices, fluorescence, phosphorescence, and chemiluminescence are used in biosensors to evaluate biological recognition.

The use of optical transducers for direct (without tagging) bacterium identification is very promising. When cells connect to receptors, these sensors can detect a minute rush in refractive index or thickness. They link direct changes in light patterns with alterations in concentration, mass, or the number of molecules. Monomode dielectric waveguides, ellipsometry, the resonant mirror, surface plasmon resonance, and the interferometer, *etc.* are also used to study the optical signals.[17–19] One significant benefit is that optical biosensors may be reduced to miniscule size utilizing fibre optics, semiconductor photo devices, and quantum dot-based sensing modules.

Fibre optics are used in optical biosensors to transform the emission signal to a recognizable fluorescent signal. Piunno *et al.* used the fluorescent marker ethidium bromide to identify a hybridization process between a DNA probe and a target. The total internal reflection was measured in the optical fibre, which was found to be proportional to the total quantity of intercalated ethidium bromide.[20]

10.2.3.5 Surface Plasmon Resonance (SPR) Biosensors

SPR is the cumulative oscillation of conduction band electrons in resonance with the oscillating electric field of incident light, which produces strong plasmon electrons through non-radiative activation. To use the SPR technique, a phonon field-based optical sensor for an optoelectronic device is made with a thin gold layer. Through the measurement of reflection minima on photodetector grid sensors, the interactions between a sample running over fixed interactant onto the gold surface is investigated. By using immunoreactions, SPR has been used effectively to detect pathogenic bacteria.[16,21]

10.2.3.6 Electrochemical Biosensors

In biosensor development, electrochemical devices have typically gained much more attention. A molecular sensing device that intricately binds a biological recognition element to an electrode transducer is referred to as an electrochemical biosensor. The transducer's job is to transform a biological recognition trigger into an electrical signal that may be used. These biosensors provide a simple, low-cost platform for patient diagnostics that is quite sensitive and specific. Electrochemical biosensors are primarily used to monitor modified DNA, DNA-binding pharmaceuticals, glucose levels, and other biological molecules. Electrochemistry enables the researcher to operate with suspended solid samples, and the initial cost of equipment is substantially lower than with optical approaches. However, on the other side, electrochemical approaches have slightly lower specificity and sensitivity than their optical counterparts. Due to high electrical conductivity, carbon nanotubes and graphene-based nanomaterials enhance electrochemical signals. These biosensors are categorized as conductimetric, amperometric, or potentiometric based on the electrical properties they measure.[19,21]

10.2.3.6.1 Conductimetric Biosensors. Here the electrical conductance/resistance of the solution is the parameter that is monitored. The overall conductivity or resistivity of the solution changes when electrochemical reactions generate ions or electrons. This shift is quantified and calibrated to a scale that is appropriate. The sensitivity of conductance measurements is however poor.[19]

10.2.3.6.2 Amperometric Biosensors. This is probably the most widely utilized electrochemical detection approach in biosensors. The great sensitivity and wide linear range in identifying electroactive species in biological samples make the amperometric biosensor more appealing. Excellent research into new sensing principles, along with a slew of technology advancements, has paved the way for broad clinical use of amperometric devices.[22,23] Amperometric biosensors detect the current associated with the reduction or oxidation of an electroactive species entangled in the character recognition by applying a constant voltage. The current produced by amperometric biosensors is proportional to the concentration of the analyte to be detected. Because of this miniaturization, many minuscule electrode transducers may be packed into a compact package of a biochip device, allowing for the creation of high-density displays. The Clark oxygen electrode is used in the majority of amperometric biosensors. Sophisticated electrical bioassays can compete with the most advanced optical methods due to their great sensitivity, specificity, simplicity, and intrinsic downsizing.[16,19]

10.2.3.6.3 Potentiometric Biosensors. The oxidation or reduction potential of an electrochemical process is recorded in this kind of sensor, which is the least common of all biosensors. The working principle relies on the fact that when a voltage is applied to an electrode dipped in a solution, a current flow occurs because of electrochemical reactions. The analytical information in potentiometric devices is derived by translating the bio-recognition activity into a potential signal in conjunction with the application of ion-selective electrodes (ISEs). The voltage at which these reactions occurs indicates a particular reaction and particular species.[16]

10.2.3.7 Bioluminescence Sensors

The capacity of certain enzymes to generate photons as a consequence of certain processes or reactions, has recently been exploited for bioanalytical sensors.

This phenomenon is referred to as bioluminescence. In the field of mycobacteriology diagnostics, luciferase reporter phages (LRPs) shown a lot of potential. Its invention sparked the prospective applications of bioluminescence for bacterial identification. The inducible or constitutive expression of the bacterial luminescence lux gene has been widely used as a reporter. The reporter lux gene is linked to a promoter that is regulated by the concentration of a chemical of interest in the inducible approach. As an outcome, by measuring the bioluminescence intensity, the concentration of the molecule may be determined quantitatively. Detection of a wide spectrum of microorganisms has been done using bioluminescence sensors.[16]

10.2.3.8 Microbial Biosensors

In the development of biosensors, microbes provide a range of potential advantages as biological sensing substrates. They can metabolize a wide spectrum of chemical substances and are found in almost every habitat.

Microorganisms have quite a remarkable ability to cope even in harsh environments and, over time, to develop the potential to decimate novel compounds. Microbes can also be genetically modified *via* mutation or recombinant DNA technologies, making them a cost-effective source of intracellular enzymes. Because of their high specific activity and analytical specificity, purified enzymes have been the most widely used in the design of biosensors. Intracellular enzymes account for more than 90% of all enzymes discovered to date. Therefore, it has been revealed that using live cells as a resource for intracellular enzymes is a viable choice in isolated enzymes in a variety of industries and applications. In the development of biosensors, living cells are becoming extremely important. Active bacteria degrade various organic substances, potentially anaerobically or aerobically, leading to a number of end products such as ammonia, carbon dioxide, acids, and other compounds that can be detected using various transducers.[24]

10.2.3.9 Nucleic Acid-based Biosensors

A nucleic acid biosensor is a device that combines an oligonucleotide with a signal transducer to carry out analysis. The nucleic acid probe is attached to the transducer and serves as a bio-recognition molecule for detecting DNA/RNA fragments as well as biological and chemical entities. An oligonucleotide with a known sequence of purines and pyrimidine bases, or with a complex framework of DNA or RNA, is used as the sensing material.[16] Among the nucleic acid-based probes, DNA is recommended as a sensing tool.

The capacity to amplify a desired target DNA from the host pathogen *via* PCR and so increase the signals produced by the biosensor is a great advantage of DNA-based probes. Unlike enzymes or antibodies, which act as detecting molecules, DNA creates easily composable bio-recognition levels that are extremely stable and reusable after simple melting of DNA duplex.[25]

10.2.3.10 Enzyme Biosensors

Enzyme biosensors are used for the purposes of clinical analysis, food standards monitoring, or ailment tracking. It is a type of analytical technique that integrates an enzyme with a transducer to generate a signal proportional to the concentration of the sample components. This signal can be caused by a change in proton level, the discharge or consumption of gases like ammonia or oxygen, photon emission, absorption or reflection, heat emissions, and so on, all of which are caused by an enzyme-catalysed process. Through electrochemical, thermal, or optical techniques, the transducer translates this signal into a quantifiable response, like current, potential, temperature change, or light absorption. These signals can be augmented, analysed, or saved for further utilization.

Now it is the time to discuss nanobiosensors briefly.

10.2.3.11 Nanobiosensors

Sensors relying on nanotechnology are known as nanosensors. Amongst the most recent developments in the field of nanotechnology is the introduction of nanobiosensors. When combined with nanoparticles, the sensing device is referred to as a nanosensor, and when biomolecules are integrated into the device, it is referred to as a nanobiosensor. This sort of biosensor is capable of detecting changes at the atomic level with greater precision. These nanobiosensors are nanomaterial-based biosensors created by integrating the expertise of several disciplines, including bioelectronics, material science, miniaturization methods, electrode designing, fabrication technology, nanolithography, microfluidics, *etc.* Silver and some other metal nanoparticles are being used in biolabelling and nanobiosensors.[26] In comparison to typical sensors, nanobiosensors allow for the rapid and accurate detection of microbes, poisons, and adulterants. Degradable dietary components, such as vitamins and antioxidant compounds, can also be detected using these sensors.

Let us begin with the applications of sensors in the food industry.

10.3 A Quick Overview of Nanotechnology

Nanotechnology refers to the use of materials with nanoscale structures ranging from 1 to 100 nm and is an inter-disciplinary field that includes physics, chemistry, biotechnology, and engineering. Materials acquire distinctive features at these nanoscale level size range that were not present when they were in their basic bulk form. Nanomaterials have different structural components, spheres, molecules, clusters, or crystallites in different dimensions, such as zero dimensions, which includes nanoparticles, nanoclusters, and quantum dots; one-dimensional nanorods and nanotubes; two-dimensional nano-thin films; and three-dimensional nanomaterials in the 1 to 100 nm size range.[27] Solubility, toxicity, strength, magnetism, diffusivity, optics, colour, and thermodynamics are among the distinctive physicochemical properties of these nanomaterials that are unmatched due to their large surface area to volume ratios. Because of their unique capacity to enhance solubility and bioavailability while processing and storing bioactive components, these nanoparticles are finding applications in a variety of fields, notably agriculture, medicine, cosmetics, food, and public health.

10.4 The Food Industry *vs.* Nanotechnology

Since nanotechnology is the study of extremely small materials, and it has the potential to have a large impact on the food market, it encompasses a wide range of applications in the food sector. For availing its applications in the food industry, a certain type of nanomaterial is integrated into a particular food product to develop specified desirable qualities. Nanotechnology has also been a key component of research and development for the sizable manufacture of agricultural products, treated meals, and beverages, in the manufacturing, packaging, storage, transportation, and value addition of food and feed around the world. Even with its submicroscopic nature, nanotechnology has a lot of potential for providing colour, flavour, and nutritional value addition to food stuffs, extending food shelf life and regulating food integrity using barcodes including in the cold chain.[28] Nanotechnology provides various advantages in the food sector, including lower prices, reduced environmental issues, medical diagnosis to minimize losses, and improved farm management practices. It has revolutionized the food industry. Nanoparticles have roles at various phases, from inception to reaching the consumer, as in crop disease tracking and prevention, smart delivery mechanisms for bioactive components, fertilizers, pesticides, and fungicides, minimal use of harmful chemicals in the food chain, coating of enzymes for different processes, use of nanosensors in food packaging and logistics, and in detecting food poisoning and adulteration.[29]

Nanotechnology concepts give a solid paradigm for better understanding of food ingredient interactions and assemblage behaviour into nanostructures, which is anticipated to alter food structure, hydrodynamics, and nutritional qualities at the nanoscopic scale. It could improve packaging systems by modifying material permeability by using synthetic nanoparticles. It is a new field of science with the potential to revolutionize the food industry with radical new products and procedures.

The food market necessitates innovative technologies in order to maintain market leadership in the food processing business and provide fresh, authentic, convenient, and tasty food products while also extending the shelf life and freshness of the product with enhanced food quality. Food and agricultural production, food processing, marketing, storing, nanoadditives, cleaning, and sensors for toxin detection, as well as the design of unique goods, are all expected to take advantage of the new substances, products, and technologies. Nanotechnology is gaining popularity in this field also and it opens up new prospects for the food marketing strategy.

Nanotechnology is expected to affect the whole food business, altering the way food is produced, processed, packaged, transported, and served. Food is subjected to various different post-harvesting and processing alterations that really are capable of changing its biological and biochemical characteristics, and nanotechnology advancements might have an impact on the food business in the future.

10.5 Where Nanomaterials are Required in the Food Industry

A food is termed nanofood when nanoparticles or nanotechnology techniques are employed during the cultivation, production, processing, or packaging of the food.

In food industry, nanofood implementations are widely recognized as additives to foods (nano inside) and packaging for food (nano outside). They are employed during food processing, manufacturing, or thermal treatments (*e.g.* coagulation, emulsifying, or homogenizing). Food proteins, like polysaccharides and lipids, are globular particles that range in size from 10 to hundreds of nanometres. As a result of coagulation and emulsification, reticular, zero-, one-, two-, and three-dimensional nanostructures are formed. When corn starch is heated, it melts into minuscule three-dimensional crystalline structures that are barely tens of nanometres thick.

Consumer concerns about food quality and health benefits are driving scientists to develop a technique to improve food quality while preserving the nutritional content of the product to the greatest extent possible. The food industry has increased its demand for nanoparticle-based products, as many of them include vital nutrients. Colouring agents and nutritional elements such as minerals and vitamins can be encapsulated in nanoparticles to increase their bioavailability and solubility, allowing for regulated release and safeguarding biologically active chemicals and micronutrients throughout processing. Nanomaterials are widely utilized against numerous pathogenic bacteria, in health care, crop protection, water treatment, food safety, and food preservation due to their outstanding physiochemical nature and antimicrobial capability.

From food production to processing to packaging, nanotechnology has holistic food solutions. Nanomaterials make a substantial improvement to food quality and safety, as well as protecting the health advantages that food provides.

Nanotechnology may enhance the effectiveness of interactive or functional foods that satisfy the needs of the human body and supply nutrients, perhaps with better efficiency. Many institutions, researchers, and corporations are developing new approaches, procedures, and products that use nanotechnology in food science directly.[30]

Research efforts are therefore working in this direction of on-demand foods that will remain latent in the body system and supply nutrients to cells only when they are needed. Thousands of nanocapsules holding flavour or colour boosters, as well as additional nutritional ingredients (such as vitamins), would lay dormant in the food and be released only when required.[31]

Nanomaterials as food additives for food processing, preservation (nanosensing), and nano-enabled packaging are three types of nanotechnology used in the food industry.

Food processing, preservation (nanosensing), and nano-enabled packaging are three categories of nanotechnological applications in the food sector where nanoparticles are used as food additives.

10.5.1 In Food Processing

The conversion of agricultural and animal products into food is known as food processing. It also encompasses a wide range of food processing techniques, from grain milling to make raw flour for domestic cooking to extensive technological innovations used to produce easy ready-made foods.

Intelligent nutrition transport, bioactive nutritious nanoencapsulation, protein bioseparation, quick monitoring of biological and chemical toxins, solubilization, distribution, and colour in food systems are all areas where food processing can be enhanced.[32]

Food processing is the practice of preserving food using procedures and processes to turn it into a consumable state. These approaches are aimed at preserving the taste, texture, and quality of the food, while also protecting it against microbial intrusions, which can lead to food degradation. Nutraceuticals, gel-forming and viscosifying agents, nutrition delivery, mineral and vitamin enrichment, flavour, and nanoencapsulation are some of the food processing technologies that use nanomaterials.[33,34]

Food processing innovations include adding nanoparticles to conventional foods to improve nutritional assimilation, such as incorporating tuna fish oil, an omega-3 fatty acid source, through nanocapsulation.[35] The role of nanotechnology in food processing can be evaluated by considering food texture, food appearance, food flavour, food nutritional value, and food shelf life.

10.5.2 In Packaging

Consumable items, such as food materials, are available only for a limited time period. At any point in the food chain, they can be polluted and/or degraded through chemical, physical, or biological processes. Any infection introduced into or present in food can cause poisoning, which is potentially fatal. Packaging is essential for maintaining the freshness of food also. Gas and moisture permeability, as well as strength and biodegradability, are required in a good packaging material.[36] Food must be safeguarded at every stage. For that, high-quality, safe, non-toxic, and non-expensive packing materials are needed. The pH, temperature, dampness, and freshness of the item inside the packet can all be maintained utilizing nanoparticles in packaging, which also provides user information and safety regulations to increase the food shelf life. Food protection systems are improved by smart packaging, ensuring that it is safe to consume as well as increasing its shelf life. Food packaging can incorporate nanomaterials as nutrient transporters, additives, or antibacterial agents. Their optoelectronic qualities can be employed as sensors and transducers to evaluate food quality, while their mechanical strengths and endurance can help improve package quality.

Metal or metal oxide-based nanocomposites are utilized in product packaging, coatings, and occasionally as food additives. In the food sector, silver nanoparticles and their nanocomposites are the most frequently exploited nanomaterials as antimicrobial agents.[37]

Thus nanomaterials are used in food industry for a variety of reasons, including:

1. Nanoencapsulation of beneficial substances, nutrients, or nutraceuticals to safeguard and improve their stability;

2. Providing protective effects by covering surfaces with active nanoparticles;
3. Fillers to increase packaging's mechanical strength;
4. Sensing elements, properly coupled and integrated in packaging, to develop methods for the *in vitro* evaluation of food safety standards.

Nanotechnology in the food industry works in two categories:

(i) Food nanostructured components and
(ii) The food nanosensing field.

Food nanostructured components are used in a variety of applications, ranging from food processing to preservation to food packaging. These nanostructures can be employed in food processing as food additives, carriers for smart nutrition delivery, anti-caking agents, antibacterial agents, fillers to enhance the mechanical strength and structural integrity of packaging materials, and so on.

Food nanosensing, on the other hand, can be used to improve food safety and quality assessments.[38]

Food sensory qualities, such as aroma and flavour, influence usage and are thus crucially significant. Sensory qualities, on the other hand, can be difficult to manage and maintain, particularly during food processing and preservation.

Nanosensors can detect and sense food contamination caused by pathogens, microbes, fungi, *etc.*

10.6 Nanosensors/Nanobiosensors: Sensors for Food Analysis Using Nanotechnology

Throughout the food manufacturing and supply chain, food safety monitoring is critical. The majority of food analysis is now done in centrally controlled labs using standard laboratory equipment, and only a small number of samples may be examined at a time. Food manufacturers, storage facilities, and customers would be able to monitor food quality and safety during shipping, storage, and consumption by using quick, economical, and easily portable handy assays. Nanomaterials have enabled the creation of a wide range of sensor devices that can evaluate food at source.

Nanosensors are nanogadgets that can sense and evaluate physical parameters by converting them into signals that can be recognized and studied. Such nanosensors are diagnosing devices that are used to assess the quality of foodstuffs as well as food safety. These are employed in foodstuffs to detect very small micro-levels of any chemical toxins, viruses, or pathogen. Nanoparticles could preferentially adhere to foodborne pathogens, and even trace amounts of hazardous bacteria could be spotted using infrared radiation or magnetic nanoparticles. Multiple nanoparticles could be placed on a single nanosensor to sense various microbes, pathogens, and diseases easily and reliably.

Combining biology and nanoscale engineering in the production of sensors (biomolecules integrated into the device) with both a high potential for increased sensitivity and possibility of shortened reaction time can be achieved is possible. This is now described as a nanobiosensor. This type of biosensor, as described earlier, can detect changes at the atomic level with increased accuracy. Biomolecules, such as sugars or proteins, may be used as specific target groups for nanostructures as biosensors.

These nanobiosensors could serve as detectors for food pathogens and other contaminants as well as can monitoring food characteristics. Nanobiosensors may play a key role in detecting pesticides, even in trace amounts, a minuscule amount of organic molecules, a minimal concentration of microbes, and other potentially hazardous chemicals in a wide range of food products, enabling an excellent alternative solution to food safety.[39]

Nanobiosensors have special advantages over other conventional methods (*e.g.* gas/liquid chromatography, mass spectrometry, *etc.*), as they have a high surface area to volume ratio, which allows for even more antibody/enzyme/reactant loading capacity due to increased sensing interface, lower validation limits, exemplary selectivity with diminutive size, and a rapid response.

When it comes to the production, processing, transportation, and storage of food, one of the most serious concerns is bacterial degradation. Newer nano-antimicrobials have shown promising potential in preventing food degradation and extending the food storage period.

Coating a nanobiosensor with a substance can cause it to glow in different colours when it interacts with pathogenic microbes found in food. For the prompt identification of the presence of *E. coli*, a hybrid nanosensor coupling magnetic resonance and fluorescence was fabricated, which is able to identify varying quantities of a pathogenic strain of *E. coli* in a milk test in less than an hour.[40]

Therapeutic efficacy has been found in different metal and metal oxide nanostructures. Their physicochemical features are found to be the cause of excessive reactive oxygen species (ROS) production, which results in oxidative stress and eventual cell death.[41]

10.6.1 Nanosensors in Packaging

Nano-natured 'smart' and 'active' food packaging materials offer a number of advantages over traditional packaging methods. Active and intelligent packaging technologies monitor changes in the food package and alert consumers, all while releasing active essential nutrients that keep the food fresh. From improved packaging material with enhanced mechanical strength, impermeability, and antimicrobial films to nanosensing for pathogen detection and notifying consumers about the safety status of food, nanocomposites can be utilized as an active material for packaging and material coating to improve food packaging.[42] Nanosensors have been designed for food monitoring and tracking purposes. Recent studies focus on the integration of nanobiosensors onto food packaging materials for monitoring, safeness, and biosecurity, as well as the design of efficient biosensors for the diagnosing and detection of pathogenic microorganisms in foodstuffs and the associated surroundings. Nanotechnology allows for the use of nanosensors in food packaging to check quality at different stages of the transportation system to ensure the quality of the product for the ultimate user.[43]

Nanosensors developed for smart packaging sense food spoiling and discharge nano-antimicrobials as needed, thus improving the shelf life. Smart packaging products and distribution, sense changes in the food packaging and inform consumers, meanwhile discharging proactive essential nutrients that keep the food fresh. This aids marketing and the preservation of food for a prolonged period of time. Because of constantly changing consumer habits, nano-enabled packaging is a growth market in the food, beverages, and pharmaceutical industries. Minor damages like crevasses posed by external factors such as temperature and dampness, *etc.* can be mended using such smart nanopackaging technologies, and the consumer also can be informed if the food is tainted for any reason.

The use of nanobiosensors in packaging also offers information about the enzymes produced during the decomposition of food molecules, which renders the food unfit for consumption. Packaging extends the food quality, safety, and shelf life by restricting the entry of oxygen and many other enzymes from the air, as well as reducing the usage of artificial preservatives. It moreover aids in the removal of the ripening hormone ethylene, thus enabling the shelf life of products to be extended. Low-cost nanosensors can also be incorporated into food packaging to detect changes in food quality during storing and transport. These sensors use visual, optical, or electrical outputs to communicate these alterations. Gases, chemical pollutants, odours, temperature and light intensity, pathogens, and microbial metabolic products are all detected using nanosensors in smart packaging materials.[44] Nanosensors based on conducting polymers are also exploited as they can identify and quantify microbes while relying upon their gas emissions (Scheme 10.1).[45]

Rutgers University researchers have proposed an electronic tongue for Kraft Foods packaging that consists of an assemblage of nanosensors that are particularly sensitive to gases emitted by decaying food. They have created sensor strips that change colour to indicate whether or not the foodstuff is fresh and healthy.[46]

Variations in foil permeability, improving barrier function (thermal, chemical, mechanical, and microbial), strengthening mechanical and heat tolerance qualities, creating effective antimicrobial and antifungal crevices, and perceiving any contaminant and pharmacological changes are all problems that nanoscience can address.[47]

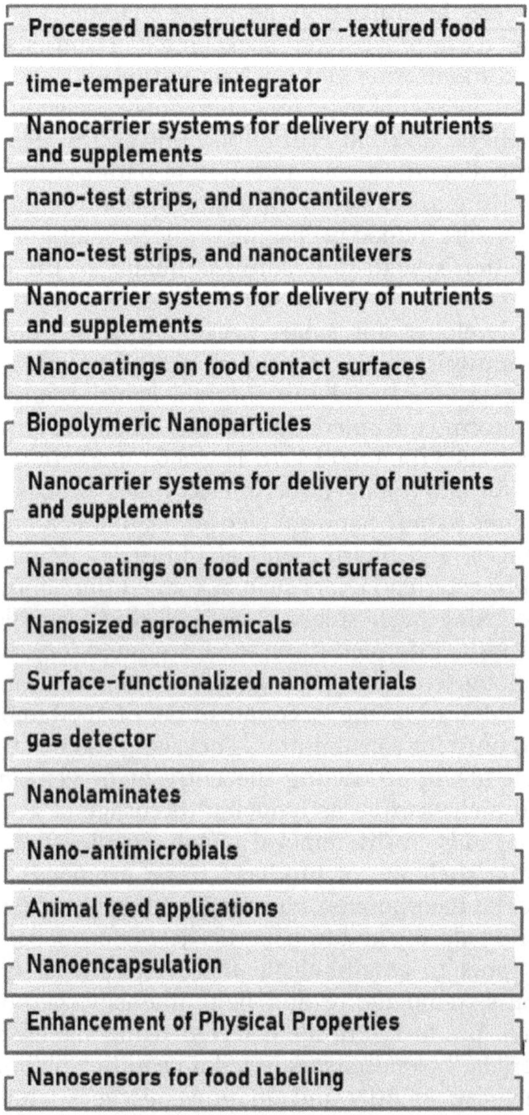

Scheme 10.1 Various implications for nanomaterials in the food industry.

10.7 Concerns about the Consequences and Safety

The functionality that can be achieved by relatively small amounts of the engineered nanoparticles, their increased interfacial reactions that improve the effectiveness of their use, and their easy formulations, better handling, and lower environmental impact are the most promising applications of nanotechnology in food technology.

Although food nanotechnology employing nanoparticles has advanced rapidly, on the other hand, nothing is known about nanoparticle toxicity. Nanomaterials have several unique qualities, such as a large surface area that literally makes them more highly chemically active than their bulkier form. As a result, they could even be involved in the majority of biochemical processes that really are hazardous to our environment and human health. We must be extremely cautious that nanostructures used in the food or associated industries do no harm, either actively or passively. Because of the special features of nanoparticles, the food and related industries are already undergoing substantial changes. However, their distinct characteristics can often have damaging consequences for entire ecosystems. Inflammation and heavy metal leakage are other major safety issues associated with nanoparticle consumption. They are currently being incorporated into food commodities at a faster pace than envisioned, with resulting health and environmental consequences. Nanomaterials must be used with caution because they have the ability to cause hazardous effects, and their usage in food technology is continuing to grow faster now than ever.

Potential uses of nanotechnology in food have been identified by food industry professionals and researchers, yet the success of any developing technology is highly dependent on its expense, budget, and public perceptions of its threats and merits. Various policy makers are analysing the safety risks associated with nanotechnology products and developing more unified and coordinated approaches to the improvised nanotechnology items.[48] With the increased use of nanoparticles, specialists' concerns have grown, but the public has remained uninterested in this issue despite increasing knowledge of nanomaterial toxicity. Several studies have found that the public general reactions to nanotechnology awareness are neutral. In general, public opinion on nanotechnology applications is mixed, ranging from indifferent to slightly positive. Consumers may want to be cautious while consuming nanofoods. Some experts[49] have investigated the general public's attitude regarding nanotechnology's acceptability in food science. They found that 30% of participants highlighted the benefits of nanotechnology to the dangers, 44% considered risks to be coincidental with benefits, and 26% believed the risks outweighed the overall benefits.

Individual preferences or choices, as well as understanding of the science behind the product's manufacture, determine consumer acceptability of any unique food product. Although nanotechnology's focus on sustainability for individual and community is commendable, there is insufficient evidence in light of the health concerns associated with its use.

Antimicrobial qualities of silver, zinc oxide, titanium oxide, and titanium nitride as nanoparticles in various packaging techniques have been much realized, including with zinc oxide being more effective and economical than most others. The antibacterial effect is attributable to the photocatalytic reactions of nanostructured particles causing bacterial cell membrane rupture.[50]

Simply select any model organism and any nanoparticle, then, on discovering the toxicological effects of the same nanomaterial, that will be either contradictory or somewhat different.

It has been just recently realized, after years of investigation, that materials at the nanoscale level exhibit significantly different characteristics and unanticipated behaviours. The concerns about its toxicity arose as a result of this unexpected behaviour.

The interactions of synthetic nanoparticles with various biological entities and the environment remain mostly unknown. The majority of the investigations published are all based on *in vivo* studies, in which a specified dose of nanomaterial is injected into cells growing at the base of the tube or plate. The behaviour of the cells is then examined in order to determine the impact that the nanomaterial has on the cells. These investigations cannot be compared because they are often carried out under random conditions and with poorly characterized nanomaterials. Very few studies consistently provide useful information. Although there has been significant development in nanotoxicology, there is, in fact, a gap between validation and conclusive data-based investigations. The optimal dose and permitted limits for substances are determined by the dose–response relationships obtained from toxicological studies. Contrary to traditional metabolites and their toxicological studies, nanoparticles have radically distinct forms, surface areas, and surface electrical charges that are different from their bulk equivalents.[51]

The degree of personal exposure to nanoparticles is highly dependent on the specific area and concentration of food products used, with the threat of exposure being higher in fields where nanomaterials are added directly to the food products as carriers in the form of creative newer food additives. The diffusion of nanoparticles from food packaging materials, for example, as well as their behaviour once they enter the body, has still to be thoroughly explored. The physiological fate of nanomaterials upon consumption, their behaviour within the gastrointestinal tract, and their potential interactions with biomolecules must be studied using the appropriate techniques. For different types of nanoparticles, different methods of action are likely to be more or less relevant. The potential of inorganic nanoparticles to be absorbed by the body, aggregate in specific tissues, and cause cytotoxicity are the most plausible processes. To consider the ability of organic nanoparticles to increase the bioavailability of potentially harmful chemicals (like pesticides or hormones) or substances that are only poisonous at high concentrations (such as certain fat-soluble vitamins) is probably more important. However, further research is needed to determine the size and significance of their impacts.

Nanoparticles can produce oxidative stress in human body cells and can migrate from the lungs to the bloodstream, cell nuclei, and the central nervous system, causing gastrointestinal inflammation, and may cross the blood–brain barrier causing Parkinson's disease, Alzheimer's disease, and DNA damage. Long-term exposure to nanoparticles may be associated with negative effects on the kidneys, liver, and other body systems also.[52]

10.8 How Ingredients Affect the Mechanism of Metabolism of Nanofood Materials: Fate of Nanoparticles in the Gastrointestinal Tract

The greater risk of nano-engineered particles is due to their higher reactivity and the increased bioavailability of smaller particles in human bodies, both of which might have long-term harmful repercussions.

Nanomaterials can enter the food chain through direct inclusion of nanoparticles in innovative foods such as nanoemulsions, nanocapsules, and nanoantimicrobial coatings, as well as nanomaterials especially used in nanosensors, nanobiosensors, and nanolaminates employed in food manufacturing, processing, preservation, and tracking. Silver, titanium dioxide, silicon dioxide, iron oxide, and zinc oxide are some examples of inorganic nanoparticles and organic nanoparticles such as lipids, proteins, and carbohydrates that are added to novel foods.

The structural composition of nanoparticles has a significant impact on how they end up in the gastrointestinal tract. Proteases, lipases, and amylases are enzymes that break down lipids, proteins, and carbohydrates in the mouth, stomach, small intestine, and colon. Some organic materials as well as inorganic ones utilized to make food nanoparticles may not be metabolized in the upper gastrointestinal track. Due to changes in pH or dilution, however, some of them may completely or partially cause the substance to dissolve.[53]

These substances may disperse, agglomerate, deposit, or even alter the physicochemical properties of the media in which they are held.

The unusual behaviour of nanomaterials and their cellular absorption are not taken into consideration in standard experiments. It is a general perception that the standard *in vitro* experiments may misinterpret the results and dose–response curves.[54] Oberdörster *et al.* found a substantial increase in inflammatory symptoms or parameters while administering 20 nm TiO_2 particles compared to the same mass of 250 nm particles in their inhalation experiments. TiO_2 particles were formerly thought to be innocuous and were used as an inert control in various toxicological tests until their experiments. Impairments to the pulmonary epithelium, blocking of Kohn's pores, the formation of interstitial fibrosis and changes in macrophage activities (pro-inflammatory factors) were all considerably higher. These findings reveal that when inert particles are nanoscaled, they can become physiologically active. The majority of toxicity studies have been undertaken at much higher doses than are reasonable.[55]

10.8.1 Dimensions/Particle Size

Nanoparticle toxicity , penetration, and permeation in the GIT are all determined by their dimensions or particle size.

To begin with, smaller nanoparticles dissolve or are digested more quickly in GIT fluids than larger ones of equal composition.

Second, due to the increase in surface area, the capacity of GIT components to interact with nanoparticles (such as digestive enzymes, phospholipids, bile salts, or mineral ions) is predicted to rise as their size reduces.

Third, as nanoparticle size is reduced in comparison to the pore size of the biopolymer network, their penetration through the mucus layer coating epithelial cells normally rises.

Fourth, particle size influences whether nanoparticles are taken up by intestinal epithelial cells *via* vascular endothelium, facilitated diffusion, or through any passive carriage processes.[51,56,57]

The majority of these contemporary toxic substances are safe enough in controlled and small doses, and only become fatal when consumed in large quantities. It is challenging to determine the precise quantity of hazardous toxins and nanoparticles released into the atmosphere. It is well said that it is the dose that makes a substance poison.

In research conducted by Chau *et al.*, it was found that 20–30% of polystyrene nanoparticles with diameters of 50 and 100 nm were absorbed by the intestinal mucosa, with smaller particles penetrating quicker.[58]

When it comes to nanoparticles in food, the particle size and mass are essential factors to consider when determining toxicity. When nanoparticles are disseminated in food products, their physicochemical and structural properties may be significantly altered, which could affect their GIT destiny and toxicity.

10.8.2 Interfacial Interactions

The fate of food-grade nanoparticles in the GI tract, and consequently their potential for severe health impacts, is frequently determined by their interfacial properties. The electrical charge, hydrophobicity, texture, digestibility, and chemical reactivity of nanoparticles in foods and the behaviour of food-grade nanoparticles are primarily controlled by their interfacial properties. Therefore, within GIT, the coating of nano-sorbent materials, characterized by their specific electrical charges, hydrophobicity, texture, digestibility, and chemical reactivity of the interface play an extremely important role, for example, their capability to pass through biological membranes such as the mucosal layer or intestinal epithelial cells as well as their interactions with other GIT elements such as mucin, digestive enzymes, bile, mineral ions, proteins, and so on.[59,60]

10.8.3 Aggregation

Food-grade nanoparticles can be found as single particles or in concentric groups with varying sizes, morphologies, and strengths. Physical factors such as van der Waals, electrostatic, hydrogen bonding, and hydrophobic forces typically hold nanoparticles together in groups. The state of aggregation of nanoparticles is thus frequently influenced by external factors such as pH, ionic strength, component interactions, and mechanical stresses. Nanoparticle clusters can have significantly larger dimensions than individual nanoparticles, which has a significant impact on their GIT destiny, such as their capacity to travel through gastric fluid, mucus, or epithelial cells. As a result, rather than determining the dimensions of the initial nanoparticles introduced to foods it is really necessary to identify the truly effective dimensions of the nanoparticles present at the point of activity. Nanoparticles first move through the mouth, which has a pH of approximately 7, and then come into contact with saliva, which contains mucin, digestive enzymes (such as amylase), and electrolytes. The nanoparticles subsequently pass down the oesophagus and then into the stomach, where they come into contact with extremely acidic gastric secretions with a pH of about 2–3, which includes digestive enzymes, gastric lipase, pepsin, *etc.* and electrolytes. The nanoparticles subsequently pass through the pylorus sphincter and into the small intestine, where they are contacted by fluids containing bile, phospholipids, digestive enzymes, pancreatic lipase, proteases, amylase, and electrolytes, with a pH of roughly 5–7. If the nanoparticles do not assimilate in the upper GIT, they will eventually reach the colon, having a pH of 6–7, where they may come into contact with colonic bacteria and undigested foods. When nanoparticles are consumed with edibles, they may be released into the GIT fluids as the food matrix breaks down and is digested. On the basis of the response of the nature and structure of the food, it is determined at which GIT region it is released in.[61–65]

Biopolymers such as proteins, polysaccharides, glycoproteins, enzymes, *etc.* can bind to nanoparticle surfaces and affect their interfacial properties, or they can change their aggregated state by encouraging or inhibiting coagulation. As a result, the characteristics of nanoparticles reaching certain locations inside the GIT may differ significantly from those of the nanoparticles consumed.

For various nanoparticles, different methods of interactions are likely to be quite relevant. The ability of inorganic nanoparticles to be taken into the body, get aggregated in specific tissues, and cause cytotoxicity are probably the most critical approach. Organic nanoparticles have the capability to increase the bioavailability of potentially harmful chemicals or molecules that are poisonous only at high concentrations are likely to be the most relevant approach.[66]

10.8.4 Nanoparticle Migratory Problems

Current research results have shown that silver nanoparticles migrate into food in concentrations below the allowed limits, although there is a paucity of safety evaluation criteria in this area. According to the findings of a study conducted by the National Packaging Products Quality Supervision and Inspection Centre, silver relocation from fresh nanosilver pouches increased in all food solutions as time and temperature increased, indicating that nanosilver particles were released *via* dual adsorptive dispersion and insertion.[67]

In addition to inorganic silver metal migration, the migration patterns of modified cellulose nanocrystals have been investigated, with the migratory level of cellulose nanocrystals in isooctane being higher in 10% ethanol but still being below the European Union's constitutional limits.[68]

The first characteristic of nanoparticles is their pulmonary deposition mechanism, in which the particles, despite their small size, are deposited throughout the pulmonary system rather than just in the alveolar region. Oberdörster[69] demonstrated unequivocally that mucociliary clearance and phagocytosis are well-documented pulmonary clearance mechanisms for micrometric particles. Nano-sized particles can infiltrate the extrapulmonary organs due to their small size. Solid particles migrate *via* the epithelial layers and nerve terminals, along the neuronal axons to the central nervous system.

The migration of nanoclay from PET beer bottles was quantified using modelling behaviour, which revealed that migration occurs only when 1 nm radius nanoparticles interact with a polymer with low dynamic viscosity. The migration behaviour of two differing sizes of TiO_2 in low-density polyethylene (LDPE) packing films was investigated. The findings revealed that 100 nm TiO_2 in LDPE film migrated more into the feed solution, with the migrating trend varying based on the TiO_2 nanoparticles' compatibility with the packaging film and food stimulating liquids. According to the results of successive thorough investigations, the migration of nanoparticles in contact with plastic packaging was found to be almost negligible.[70]

The drawbacks and limitations in detection and quantification approaches are significant in migration estimations.

Although both inductively coupled plasma (ICP)-atomic emission spectroscopy and ICP-mass spectrometry are generally used to identify nanoparticles, these approaches also have their own limitations. Hence the development of an effective separation, characterization, and quantification step is critical for accurate measurement of nanoparticles.[71]

Blood circulation is boosted dramatically when nanoscale hydrophilic and positively charged particles enter the bloodstream. These nanoparticles could endanger all organs, however long-term exposure studies are confined to unknown repercussions that need to be investigated further.[72] Smaller particulates of 4 μm go deeper into the alveolar region than larger particles, causing toxicological effects once ingested. Once nanoparticles enter the body, they can pass through the blood–brain barrier, causing pulmonary granuloma, oxidative damage, and pneumonia. Nevertheless, a broad statement on the toxicity of all nanosized particles could be made through several research trials that have already been conducted.

When nanoparticles are utilized in a packaging matrix, a model developed by Simon *et al.* predicts that when the polymer dynamic viscosity and nanoparticle size decrease, the migration of nanomaterial from the packaging film increases.[73]

During their production and use, nanomaterials continue to pose a risk of entering the food chain through the air, water, and soil, causing DNA damage, cell membrane rupture, and cell death. There have been very few *in vivo* research studies on the effects of nanofoods on human and animal health so far.

Many investigations have been conducted to determine whether nanoparticles are hazardous or harmless. It has been discovered that the majority of studies on nanoparticle toxicity have solely used homogeneous nanomaterials. Because one of the key characteristics of nanoparticles is their increased reactivity, it is possible that when a benign nanoparticle is integrated in food, it may change to a risky form or *vice versa*. Food serves a variety of functions in the body, and the composition of the food plays a crucial role in those functions.

A functional component may be found in a food that is unique to that foodstuff. The primary objective of these kinds of food processing is to minimize the loss of all such beneficial components. Food, by its very nature, is a vast reservoir of biochemical interactions, and the addition of a highly reactive species of nanoparticles to food could set off a chain of interactive reactions. Numerous studies have revealed that different nanoparticles, such as silver, gold, zinc, chitosan, platinum, iron, copper, and carbon nanotubes, have synergistic antibacterial properties when combined with natural derivatives.[74]

For that, investigations have to be done on the interaction of nanoparticles with functional components and other substances.

10.9 Toxicokinetics of Various Nanoparticles

The toxicokinetics of various nanoparticles are described in Tables 10.1–10.4.

10.10 Rising Issues and Probable Remedies

Nanotechnology is making in-roads and emerging with a head of steam, and it has thus become a vital resource for the food and bioprocessing industries in meeting the requirements of the growing global population. It has almost changed the food science and business. Nanoparticles have been found to increase food quality and safety by boosting the effectiveness of food packages' storage stability and nutrient content, all without changing the flavour or physical qualities of the food. The importance of naturalness and trustworthiness in consumables has influenced people's perceptions of risk and reward.

The advancements in inorganic nanosubstance and micro-fluid production has enabled the implementation of suitable and competent sensors and biosensors to promptly identify infections, microorganisms, and insecticides. As a flavour and nutrient transporter, and as functionalized foods, nanomaterials may enhance nutritional values and preserve conventional food materials. Information is very desperately needed for a better understanding of nanoparticle–biological interactions and processes. These associations primarily involve biomolecules as proteins, although there are

Table 10.1 Toxicokinetics of inorganic nanoparticles.[a]

Absorption	Inhalation exposure and effects	Ingestion exposure	Effects on the gastrointestinal system	Cellular effects
Hussain *et al.*[75] showed cell capture of microparticulate substances by enterocytes, and their transport between cells. In some cases, the passage of microparticles from the intestinal lumen to the bloodstream led to distribution of substances in the body.	The first studies on the comparison of inhalation of fine (250 nm) and ultrafine (20 nm) TiO_2 particles found greater pulmonary retention of ultrafine particles.[55] There was a significant increase in inflammation signs during administration of 20 nm TiO_2 particles in comparison with the same mass of 250 nm particles. Particles (<50 nm) of titanium oxide, aluminium oxide and carbon black increased the pulmonary inflammation parameters 10 fold.[76] Borm *et al.*[77] pointed out that low solubility particles, such as carbon black and titanium oxide, are recognized to cause fibroses, neoplastic lesions, and pulmonary tumours in rats. Kreyling *et al.*[78] observed radioactive iridium in the animals' liver, heart, and brain. This phenomenon was twice as great for 15 nm nanoparticles.	Hillyer and Albrecht[79] reported blood and tissue distribution of ingested colloidal gold nanoparticles in mice. They noted absorption in the animals' brain, lungs, heart, kidneys, intestines, stomach, liver, and spleen, that was more pronounced for 4 and 10 nm nanoparticles.	Gatti[80] showed a correlation of the presence of microparticles or nanoparticles with colon cancer and Cröhn's disease, an inflammatory intestine disease.	An *in vitro* study by Lucarelli *et al.*[81] showed that SiO_2 and cobalt (Co) nanoparticles exhibited significant proinflammatory activity for the activity of human marrow monocytes. The Co nanoparticles showed high proinflammatory and cytotoxic potential, while the SiO_2 nanoparticles had low proinflammatory potential and the TiO_2 nanoparticles had even lower potential, although still observable, despite the fact that this substance is often considered biologically amorphous. Germain *et al.*[82] compared the *in vitro* cytotoxicity of nanoparticles of a cobalt chromium (Co/Cr) alloy and alumina ceramic, which were produced by simulated wear of prosthetic joints. Barlow *et al.*[83] exposed bovine foetal serum to carbon black fine particles (260 nm in diameter; 10 mg mL^{-1}) and carbon black nanoparticles.

[a]Insoluble inorganic nanoparticles can be composed of pure metals or various inorganic compounds including oxides or alloys.

Table 10.2 Toxicokinetics of fullerene (distributed throughout the embryo and the yolk sac).

Carcinogenic effects	Kidney effects	Development effects	Cellular and humoral effects
Slight increase in ornithine decarboxylase activity (enzyme with a role in the promotion of tumours) was noted in the epidermis.[84]	Acute and subacute toxicity of C_{60} polyalkylsulphonate in rats.[85] Several effects were reported in a 12-day subacute toxicity study by intraperitoneal injection, a significant decrease in body weight and in the weight of certain organs (thymus and heart), an increase in the weight of the spleen, and a significant rise of certain biochemical blood parameters (significant increase in aspartate aminotransferase and a significant decrease in triacetylglycerol).	The presence of C_{60} fullerenes solubilized with polyvinyl pyrrolidone inhibited cellular differentiation and proliferation of mesencephalic cells *in vitro*. Intraperitoneal administration on the 11th day of gestation caused 100% mortality and body flexion anomalies at 137 $mg\,kg^{-1}$, malformations (head and tail region) at 50 $mg\,kg^{-1}$ and increased head volume at 25 $mg\,kg^{-1}$. At 50 $mg\,kg^{-1}$, C_{60} was distributed throughout the embryo and the yolk sac was impaired. C_{60} passes through the placental barrier, disrupts the yolk sac, and causes intrauterine mortality and malformations.[86]	*In vitro* exposure to the C_{60} fullerene (12.5 μg C_{60}-cyclodextin) induced oxidative damage in rat hepatic microsomes. This damage can be modulated by antioxidants and free radical scavengers.[87] Photoinduced (halogen lamp) cytotoxicity of fullerenes has been reported in several studies.

Table 10.3 Toxicokinetics of CNTs.

Ingestion exposure	Exposure by other routes	*In vitro*	Effects on the respiratory system	Effects on the skin and mucous membranes
Hydroxylated single-walled carbon nanotubes (SWCNTs) administered by gavage in mice (100 μL of a 15 $\mu g\,mL^{-1}$ solution) are distributed to most of the organs and tissues, except the brain.[88]	The radiomarked hydroxylated singled-walled carbon nanotubes administered intraperitoneally (100 μL of a 15 $\mu g\,mL^{-1}$ solution) are distributed throughout the body, except the brain, pass through several compartments and are retained in the bones.[88]	The intracellular transport of functionalized SWCNTs, *i.e.*, conjugated with lysine, on human and mouse fibroblasts *in vitro* (1, 5, and 10 mM). showed that carbon nanotubes could pass through the cellular membrane, accumulate in the cell, and end up in the cell nucleus.[89] Cherukuri *et al.*[90] showed carbon nanotubes could be ingested by mouse peritoneal macrophages. Monteiro-Riviere *et al.*[91] found MWCNTs in the cytoplasmic vacuoles of human epidermal keratocytes (up to 3.6 μm long), a decrease in cell viability and a significant increase in an inflammation marker (interleukin-8). This demonstrates the capability of MWCNTs to penetrate the cell membrane.	The pulmonary toxicity of acute exposure to a SWCNT preparation in male rats (single intratracheal instillation; 0, 1, and 5 $mg\,kg^{-1}$). At 5 $mg\,kg^{-1}$, they reported a high mortality rate (~15%) caused by mechanical blockage of the upper airway, an increase in pulmonary cell proliferation, and an increase in multifocal pulmonary granulomas.[92]	SWCNTs caused a significant dose–response reduction of cell viability and oxidative stress biomarkers (*e.g.*, antioxidant reserve), and a significant increase in lipid peroxides in human epidermal keratinocytes.[93]

Table 10.4 Toxicokinetics of organic nanoparticles, quantum dots, and nanocapsules, -spheres and -shells.

Organic nanoparticles (genotoxic effects)	Quantum dots (effects on the humoral and cellular levels)	Nano-capsules, -spheres and -shells (insoluble organic polymers)
Leong-Morgenthaler *et al.*[94] revealed that benzo(a)pyrene dissolved in sunflower oil and encapsulated in lipid nanoparticles had the same *in vitro* carcinogenic effect on human cells as benzopyrene dissolved in dimethyl sulphoxide.	Because of their optical qualities and ability to make covalent connections with peptides, antibodies, nucleic acids, and other low-weight molecules, quantum dots are utilized as fluorescent probes in diagnostic medical imaging and therapies.[95] Chan and Nie[96] were the first to demonstrate *in vivo* that CdSe/ZnS quantum dots coated with mercaptoacetic acid could attach to blood transferrine, as quoted by Smith *et al.*[95] Cancer cells specifically absorbed this fluorescent combination. Derfus *et al.*[97] investigated the cytotoxicity of CdSe quantum in an *in vitro* investigation. The released Cd^{2+} ions have been identified as carcinogenic in humans. Green and Howman[98] in an *in vitro* experiment, found the quantum dots changed the DNA by creating SO_2 free radicals as a result of ZnS oxidation.	Cahouet *et al.*[99] administered nanocapsules (20–100 nm) with a lipid core and a shell made of 2-hydroxy-polyethylene glycol (PEG) stearate and lecithin intravenously with iodine-125 and technetium-99 label. The nanocapsules remained in the blood compartment for a longer period of time than expected due to the PEG coating. The nanocapsules were found in the liver, intestines, stomach, and penis of the animals, although there was no substantial dispersion in the brain. Zhou *et al.*[100] demonstrated in an *in vitro* investigation using a nanosphere formulation to give arsenic trioxide minimized the blood toxicity of this product, which is used to treat bladder cancer, and makes its effect more selective to cancer cells.

researches that indicate the various ways to track this. Granick focused on nanoparticles' physical interactions and discovered that these can affect the lipid membrane phase structure.[101]

In the coming years, nanotechnology-derived food products are projected to become more readily available to users around the globe. The physiochemical and skeletal/morphological features of nanoparticles found in foods and drinks vary significantly, influencing the GIT destiny and cytotoxic potential. As a result, appropriate analytical tools are needed to characterize nanoparticle properties, which previously have been discussed.[102,103]

It is crucial to discover how harmful nanoparticles are, as well as the potential environmental and health risks they bring. More research based on this concept will aid our understanding of the mechanism underlying nanomaterial-induced toxicity, as well as contribute to the development of nanotechnology that is both safe and profitable.

Furthermore, developing restrictions and regulations to safeguard society from any nanotechnology-related negative impacts is essential.[104]

Nanotoxicological studies are carried out to anticipate the unknown repercussions of nanoparticle use. In nano-toxicological investigations, a typical toxicity test involves cells or species being exposed to a precise amount of nanomaterials.

Nanoparticle toxicity has been studied using methods comparable to those used in traditional toxicity testing.

The majority of research has suggested that oxidative stress is an important criterion to consider when analysing nanotoxicity. However, while assessing toxicity, one must consider not just the nanoparticle but also a variety of several other variables. CNTs, for example, can be SWCNTs or MWCNTs, functionalized or unfunctionalized, conjugated or nonconjugated with any metal catalyst, hydrophobic or hydrophilic depending on the specific functional group attached, and so many other variables must be taken into account. Toxicologists have discovered critical characteristics or tests that allow scientists to screen safer nanomaterials because testing each and every parameter for toxicity would be too time consuming.[105]

By knowing and understanding how nanoparticles interact with biomolecules or living organisms on a physicochemical level can provide a plethora of information to toxicologists. The best parameter for distinguishing between harmful and benign materials appears to be the concept and mechanisms of oxidative stress.

Xia *et al.* made great strides forward by investigating how nanomaterials might cause oxidative damage in cells.

The goal of the research was to discover how nanoparticles work and then compare their toxicity potential. This was accomplished by simply observing the production of reactive species within the cell. The imbalance between oxidants and antioxidants leads to oxidative stress.[106]

We now can make comparative oxidative stress signatures of common particles to nanoparticles with the earlier data and these can be further extrapolated to newer ones.[107]

Although this study sheds light on the oxidative damage which certain nanomaterials can produce, it leaves a few questions unanswered. To ensure that the cell culture model is flawless, studies should be conducted on more than one control cell. Additionally, a primary cell culture ought to be employed. Furthermore, several dose administrations should be examined, with doses comparable to the dose of the nanomaterial discharged. Following that, various media formulations must be investigated, as the majority of studies have used serum with a high concentration of antioxidants. These antioxidants may be able to conceal the oxidative damage that nanomaterials induce. Therefore, some additional tests have to be done, and both intracellular as well as extracellular oxidative stress must be evaluated.[108]

Toxic doses for nanoparticles are often given in mass units. This concept, however, is no longer adequate for nano-sized particles as it does not take into account the number of particles or the specific surface area for a given mass of particles.[69] These two characteristics can be extremely high for nanoparticles and have a direct impact on their toxicity.

It is preferable to specify the dose based on the features of the nanoparticles (number and surface area) rather than their mass. Several investigations have clearly demonstrated that the hazardous effects assessed are directly connected to the surface of the nanoparticles used, rather than their mass.

Most modern toxicology studies are animal based, *in vitro* or *in silico.* There is a need for evidence-based toxicology. With colossal expertise and advancements, the conventional toxicological assays need to be redesigned and new tools, such as proteomics, functional genomics, high-throughput screening, metabolomics, *etc.* should be used more frequently.

High-throughput screening technologies and computational approaches can quickly screen and prioritize nanoparticles for toxicological experiments. At the same time, they speed up the process of establishing a link between a material's biological behaviour. In such a way the cytotoxicity of a variety of metal nanoparticles can be predicted using a quantitative structure–activity connection.[107,109,110]

To fully understand the hazards of nanoparticles in the food industry, advancements in at least three areas are required: methodology, characterization techniques, and nanotoxicity assessments. Because of the unique features of nanomaterials, procedural methods must first be devised. In these circumstances, traditional approaches will not work. In the case of nanomaterials, experts feel that different methodologies for cytotoxic studies are required. A comprehensive characterization technique is prerequisite to tracking the fate of nanoparticles in humans and laboratory animals. Recently, specific analytical methods for nanotoxicity evaluations have been developed.[111]

In terms of characterization and assays, the main cause of inconsistent information on nanoparticle toxicity may be a lack of consistency.[112] For that, standard risk assessment processes must be established. Additionally, the differences between humans and experimental animals preclude conclusions from being extrapolated.[113]

Nanotechnology packaging for food ingredients was regarded to be less challenging than nanotechnology utilized in food processing and sensing in the food business. For the assessment of hidden consequences associated with human exposure to nanoparticles, new methodologies and standardized test procedures to evaluate the effects of nanoparticles on living cells are urgently needed.

Another current problem with nanoparticles in food and related industries is the lack of a consistent and systematic taxonomy of the nanomaterials used. The process of nanoparticle manufacture and synthesis in food products must be classified and documented as well. Consumers are cautious about using nanoproducts because of the lack of such categorization. Researchers must also create appropriate *in vitro* and *in vivo* assays for toxicity testing of nanoparticles utilized in food and related industrial sectors.[114]

From the consumer's point of view, food nanotechnology still has a number of issues of concern, the most serious of which is the uncertain behaviour of nanoparticles in the body and their potential hazardous consequent effects. To mitigate such consequences, a set of standards and regulations governing their food safety implications must be devised.

For the safe use of nanoparticles in the food industry, effective rules and policies are needed. With the assistance of various international organizations, issues connected to consumer health are being regulated by modern food legislation, which can also be applied to nanotechnology and nanomaterials used in food and food products. The safety of nanoparticles in the food sector can be ensured by adhering to the current food standards, with new particular nanotechnology rules having greater visibility, and there should be a readiness to offer relevant information. To assure adoption and use of nanotechnology's manifold merits, efforts are being made to encourage international cooperation and data exchange.[115]

As a result, agencies all around the world are collecting data in order to determine the best course of action.[116]

In the United States, the USFDA is in charge of regulating nanofoods and food packaging. The Food Standards Code's regulatory agency, Food Standards Australia and New Zealand (FSANZ), plays an active role in the regulation of nanofood additives and ingredients in Australia.[117,118]

The Scientific Committee on Emerging and Newly Identified Health Risks (SCENIHR) of the European Union assesses the risks associated with nanotechnology.

According to European Union regulations, nanotechnology-based food additives must be subjected to safety testing before being approved for human consumption. The Novel Foods Regulation of the European Union covers all nanofoods and food ingredients (EC 258-97). According to the European Food Safety Authority's (EFSA) re-evaluation programme, permitted nanoadditives used prior to 2009, as well as food packaging materials, should be addressed accordingly. Whereas Japan and China are the leading producers of nanomaterials, they lack comprehensive nanotechnology-specific rules. Food regulations are lacking in various nations due to the lack of information about human exposure, accessibility, safety, and toxicity.[119] Several governments have required a regulatory structure to deal with these challenges linked with nanofood as a result of the rising regulatory issues. For legitimate nanotechnological applications, comprehensive government rules and regulations are required, as well as thorough toxicological prevention measures. For regulating the use of nanoparticles in the food sector, a largely approved international legal framework is urgently required.

Almost all governments and industries have multiple challenges in assuring public trust and acceptability of nanotechnology-based foodstuffs on the market.

Nanoparticles are produced all around the world, but only a few countries have proper legal guidelines for the use of nanotechnology in food. Limited scientific research on nanosystems makes it difficult to draw any inferences about their viability. The use of nanoparticles in food packaging is less hazardous than the use of nanoparticles as a food additive.

Adequate labelling and restrictions for the commercialization of nanofoods should be recommended in order to promote customer acceptance.

Thus, if properly managed and regulated, these nanotechnological innovations could play a key role in enhancing food processing, product quality and packaging, while promoting the health of individuals and their quality of life.

Despite the current lack of formal nanotechnology regulations and control measures, tremendous progress has been achieved in the application and regulation of innovative nanotechnology in the food industry.

The safety of nanoparticles in the food sector can be maintained by keeping current food standards, any new specific nanotechnology regulation, data governance, and a commitment to providing consumers with relevant data.

To establish the influence of nanomaterials, it is important to standardize test protocols. Additionally, policies that reduce customer anxiety and promote consumer acceptability must be implemented.

Concerns about nanoparticles in the food industry include safety, environmental, ethical, policy, and regulatory challenges. The greatest source of concern regarding nano-sized materials' possible effects and repercussions on human health and the environment is a lack of public awareness. The difficulties in identifying the diverse nanomaterials used in food, their detection and quantification in food and biological systems, inadequate information on toxicity and toxicokinetics, and a lack of optimal test methodologies makes vulnerability assessment of nanotechnologies more challenging. Whilst formulating best strategies to ensure sustainable development of nanotechnology in the food sector and regulating the potential risk to humans and the environment, legislation with detailed life cycle assessments and positive ratified procedures with particular regard to solving consumer involvement for diverse stakeholders should really be incorporated. The potential dangers of nanoparticles to human health and the environment have been now recognized. It is required to have strict quality control of nanotechnology-enabled products, as well as regulatory conformance and customer safeguarding guarantees. Though these main concepts of the emerging risk index could be implemented for nano-based foodstuffs, there have been some unresolved questions in clarify the concept the oral bioavailability of nanoparticles, the body's ability to handle them, the methodologies which could be used to evaluate their toxic effects, and in specific the statistics that could be used to estimate vulnerability.

Each use of nanotechnology in foodstuffs production, as well as the resulting effects of nanotechnology for the food web, must be evaluated on an individualized basis. As a matter of fact, risk assessment strategies for foods derived from nanotechnology will have to be extremely vigilant.

- Existing threat assessment methodologies are hampered by a variety of obstacles and restrictions that prevent them from being applied to the use of nanotechnology in the food industry.
- There are few reliable methods for detecting and measuring nanoparticles in foods and other biotic components.
- As a corollary, assessing nanoparticle consumption and ensuring various meteorological tracking are difficult.
- There are various theories about nanoparticle bioactivity and potential toxicity, as well as how they work at the molecular level.
- There are no established standard procedures for measuring nanoparticle toxicity *in vitro* and *in vivo*.
- The persistence and unpredictable nature of nanoparticles in different diets, as well as any potential interactions with food ingredients is not certain.

At the global level, regulatory mechanisms should be implemented to ensure that foodstuffs developed with nanotechnology are closely monitored to ensure that traits are reformed to the nanoscale and that they are reviewed in terms of safety.

References

1. M. M. Berekaa, *Int. J. Curr. Microbiol. Appl. Sci.*, 2015, **4**, 345.
2. F. W. H. Kampers, *Bionanotechnology: Global Prospects*, CRC Press, Boca, Raton, 1st edn, 2008.
3. D. R. Franz, P. B. Jahrling, A. M. Friedlander, D. J. McClain, D. L. Hoover, W. R. Bryne, J. A. Avlin, G. W. Christopher and E. M. Eitzen, *J. Am. Med. Assoc.*, 1997, **278**, 399.
4. R. J. Leggiadro, *Infect. Control Hosp. Epidemiol.*, 2000, **21**, 53.
5. S. Zhu, C. Dua and Y. Fu, *Opt. Mater.*, 2009, **31**, 1608.
6. L. Vermeiren, F. Devlieghere, M. Beest, N. Kruijf and J. Debevere, *Trends Food Sci. Technol.*, 1999, **10**, 77.
7. S. Banu, S. Birtwell, Y. Chen, G. Galitonov, H. Morgan and N. Zheludev, *Opt. Express*, 2006, **14**, 1382.
8. S. W. Birtwell, G. S. Galitonov, H. Morgan and N. I. Zheludev, *Opt. Commun.*, 2008, **281**, 1789.
9. R. H. Farahi, A. Passian, L. Tetard and T. Thundat, *ACS Nano*, 2012, **6**, 4548.
10. D. J. McClements and H. Xiao, *npj Sci. Food*, 2017, **1**, 6.
11. D. J. McClements and J. Rao, *Crit. Rev. Food Sci. Nutr.*, 2011, **51**(4), 285.

12. P. N. Ezhilarasi, P. Karthik, N. Chhanwal and C. Anandharamakrishnan, *Food Bioprocess Technol.*, 2013, **6**, 628.
13. S. F. D'Souza, *Biosens. Bioelectron.*, 2001, **16**, 337.
14. *Biosensors: Fundamentals and Applications*, ed. A. P. F. Turner, I. Karube and G. S. Wilson, Mir Publishers, Moscow, 1992.
15. S. F. D'Souza, *Curr. Sci.*, 1999, 77, 69.
16. R. Syam, K. Justin Davis, M. D. Pratheesh, R. Anoopraj and B. Surej Joseph, *VETSCAN*, 2012, 7, 14.
17. S. P. J. Higson, S. M. Reddy and P. M. Vadgama, *Eng. Sci. Educ. J.*, 1994, **3**, 41.
18. H. J. Watts, C. R. Lowe and D. V. Pollard-Knight, *Anal. Chem.*, 1994, **66**, 2465.
19. O. Lazcka, F. J. Del Campo and F. X. Munoz, *Biosens. Bioelectron.*, 2007, **22**, 1205.
20. P. A. E. Piunno, U. J. Krull, R. H. E. Hudson, M. J. Damha and H. Cohen, *Anal. Chem.*, 1995, **67**, 2635.
21. J. Homola, S. S. Yee and G. Gauglite, *Sens. Actuators, B*, 1999, **54**, 3.
22. J. Wang, G. Rivas, X. Cai, E. Palecek, P. Nielsen, H. Shiraishi, N. Dontha, D. Luo, C. Parrado, M. Chicharro, P. Farias and F. S. Valera, *Anal. Chim. Acta*, 1997, **347**, 1.
23. J. Wang, *J. Pharm. Biomed. Anal.*, 1999, **19**, 53.
24. R. K. Darsanaki, A. Azizzadeh, M. Nourbakhsh, G. Raeisi and M. A. Aliabad, *J. Biol. Today's World*, 2013, **2**, 53.
25. S. R. Mikkelsen, Electrochemical biosensors for DNA sequence detection, *Electroanalysis*, 1996, **8**, 15.
26. M. Rai, A. Gade, S. Gaikwad, P. D. Marcato and N. Duran, Biomedical Applications of Nanobiosensors: the State-of-the-Art, *J. Braz. Chem. Soc.*, 2012, **23**, 14–24.
27. S. S. Sanjay and A. C. Pandey, *Advanced Structured Materials*, Springer, India Pvt. Ltd., 2017, vol. 62.
28. I. E. Aigbogun, S. S. D. Mohammed, A. A. Orukotan and J. D. Tanko, *J. Adv. Microbiol.*, 2017, 7, 1.
29. B. Srilatha, *J. Nanomed. Nanotechnol.*, 2011, **02**, 1.
30. N. Dasgupta, S. Ranjan, D. Mundekkad, C. Ramalingam, R. Shanker and A. Kumar, *Food Res. Int.*, 2015, **69**, 381.
31. A. A. Amin, M. F. Salama, E. I. Seliem and K. F. Mahmoud, *Int. J. Curr. Microbiol. App. Sci.*, 2015, **4**, 1017.
32. R. Ravichandran, *Int. J. Green Nanotechnol. Phys. Chem.*, 2010, **1**, 72.
33. Q. Huang, H. Yu and Q. Ru, *J. Food Sci.*, 2010, **75**, R50.
34. J. Weiss, P. Takhistov and D. J. McClements, *J. Food Sci.*, 2006, **71**, R107.
35. B. S. Sekhon, *Nanotechnol., Sci. Appl.*, 2010, **3**, 1.
36. L. M. Couch, M. Wien, J. L. Brown and P. Davidson, *Agro. Food Ind. Hitech.*, 2016, **27**, 36.
37. X. He and H. M. Hwang, *J. Food Drug Anal.*, 2016, **24**, 671.
38. P. N. Ezhilarasi, P. Karthik, N. Chhanwal and C. Anandharamakrishnan, *Food Bioprocess Technol.*, 2013, **6**, 628.
39. S. Otles and B. Yalcin, *LogForum*, 2010, **6**, 67.

40. T. Banerjee, S. Sulthana, T. Shelby, B. Heckert, J. Jewell, K. Woody and V. Karimnia, *et al.*, *ACS Infect. Dis.*, 2016, **2**, 667.
41. H. H. Wu, J. J. Yin, W. G. Wamer, M. Y. Zeng and Y. M. Lo, *J. Food Drug Anal.*, 2014, **22**, 86.
42. S. D. F. Mihindukulasuriya and L. T. Lim, *Trends Food Sci. Technol.*, 2014, **40**, 149.
43. S. Neethirajan and D. S. Jayas, *Food Bioprocess Technol.*, 2011, **4**, 39.
44. T. King, M. J. Osmond-McLeod and L. L. Duffy, *Trends Food Sci. Technol.*, 2018, **72**, 62.
45. T. Ahuja, I. A. Mir and D. Kumar Rajesh, *Biomaterials*, 2007, **28**, 791.
46. World Bank, *ICT in agriculture (updated edition)*, World Bank Publications, Washington, DC, 2017.
47. S. M. Alfadul and A. A. Elneshwy, *Afr. J. Food, Agric., Nutr. Dev.*, 2010, **10**, 2719.
48. T. V. Duncan, *J. Colloid Interface Sci.*, 2011, **363**, 1.
49. A. H. Matin, E. Goddard, F. Vandermoere, S. Blanchemanche, A. Bieberstein, S. Marette and J. Roosen, *Int. J. Consum. Stud.*, 2012, **36**, 149.
50. S. Neethirajan and D. S. Jayas, *Food Bioprocess. Technol.*, 2011, **4**, 39.
51. S. S. Sanjay, *Nanoparticles in Medicine*, Springer Nature Singapore Pte Ltd., 2020.
52. J. Momin, C. Jayakumar and J. Prajapati, *Emirates J. Food Agric.*, 2013, **25**, 10.
53. S. Dekkers, *et al.*, *Nanotoxicology*, 2011, **5**, 393.
54. J. G. Teeguarden, P. M. Hinderliter, G. Orr, B. D. Thrall and J. G. Pounds, *Toxicol. Sci*, 2007, **95**, 300.
55. G. Oberdörster, J. Ferin and B. E. Lehnert, *Environ. Health Perspect.*, 1994, **102**, 173.
56. L. M. Ensign, R. Cone and J. Hanes, *Adv. Drug Delivery Rev.*, 2012, **64**, 557.
57. E. Frohlich and E. Roblegg, *Toxicology*, 2012, **291**, 10.
58. C. F. Chau, S. H. Wu and G. C. Yen, *Trends Food Sci. Technol.*, 2007, **18**, 269.
59. B. A. Magnuson, T. S. Jonaitis and J. W. Card, *J. Food Sci.*, 2011, **76**, R126.
60. J. J. Powell, N. Faria, E. Thomas-McKay and L. C. Pele, *J. Autoimmun.*, 2010, **34**, J226.
61. T. B. J. Blijdenstein, E. van der Linden, T. van Vliet and G. A. van Aken, *Langmuir*, 2004, **20**, 11321.
62. E. H. A. de Hoog, J. F. Prinz, L. Huntjens, D. M. Dresselhuis and G. A. van Aken, *J. Food Sci.*, 2006, **71**, E337.
63. E. Silletti, M. H. Vingerhoeds, W. Norde and G. A. Van Aken, *Food Hydrocolloids*, 2007, **21**, 596.
64. M. H. Vingerhoeds, T. B. J. Blijdenstein, F. D. Zoet and G. A. van Aken, *Food Hydrocolloids*, 2005, **19**, 915.
65. M. Mahmoudi, *et al.*, *Chem. Rev.*, 2011, **111**, 5610.
66. D. J. McClements and H. Xiao, *npj Sci. Food*, 2017, **1**, 6.

67. Y. Huang, S. Chen, X. Bing, C. Gao, T. Wang and B. Yuan, *Packag. Technol. Sci.*, 2011, **24**, 291.
68. M. Farhoodi, S. M. Mousavi, R. Sotudeh-Gharebagh, Z. Emam- Djomeh and A. Oromiehie, *Packag. Technol. Sci.*, 2014, **27**, 161.
69. G. Oberdörster, E. Oberdörster and J. Oberdörster, *Environ. Health Perspect.*, 2005, **113**, 823.
70. J. Bott, A. Störmer, G. Wolz, R. Franz, Studies on the migration of titanium nitride nanoparticles in polymers, Poster presentation at the 5th International Symposium on Food Packaging, Berlin, Fraunhofer Institute for Process Engineering and Packaging, Berlin, Nov. 14–16, 2012.
71. R. Ravichandran, *Int. J. Green Nanotechnol. Phys. Chem.*, 2010, **1**, P72.
72. C. Silvestre, D. Duraccio and S. Cimmino, *Prog. Polym. Sci.*, 2011, **36**, 1766.
73. M. Cushen, J. Kerry, M. Morris, M. Cruz-Romero and E. Cummins, *Trends Food Sci. Technol.*, 2012, **24**, 30.
74. M. Rai, P. S. Jogee and A. P. Ingle, *Int. J. Food Sci. Nutr.*, 2015, **66**, 363.
75. N. Hussain, V. Jaitley and A. T. Florence, *Adv. Drug Delivery Rev.*, 2001, **50**, 107.
76. K. Donaldson, V. Stone, A. Clouter, L. Renwick and W. MacNee, *Occup. Environ. Med.*, 2001, **58**, 211.
77. P. J. A. Borm, Hazards and risks of nanomaterials: a look forward, Compte-rendu du First International Symposium on Occupational Health Implications of Nanomaterials, Buxton, Great-Britain, 12 to 14 October 2004.
78. W. G. Kreyling, M. Semmler, F. Erbe, P. Mayer, S. Takenaka and H. Schultz, *et al.*, *Toxicol. Environ. Health*, 2002, **65**, 1513.
79. J. F. Hillyer and R. M. Albrecht, *J. Pharm. Sci.*, 2001, **90**, 1927.
80. A. M. Gatti, *Biomaterials*, 2004, **25**, 385.
81. M. Lucarelli, A. M. Gatti, G. Savarino, P. Quattroni, L. Martinelli, E. Monari, D. Boraschi, D. E. Woolley and L. C. Tetlow, *Eur. Cytokine Network*, 2004, **15**, 339.
82. M. A. Germain, A. Hatton, S. Williams, J. B. Matthews, M. H. Stone, J. Fisher and E. Ingham, *Biomaterials*, 2003, **24**, 469.
83. P. G. Barlow, K. Donaldson, J. MacCallum, A. Clouter and V. Stone, *Toxicol. Lett.*, 2005, **155**, 397.
84. M. A. Nelson, F. E. Domann, G. T. Bowden, S. B. Hooser, Q. Fernando and D. E. Carter, *Toxicol. Ind. Health*, 1993, **9**, 623.
85. H. H. Chen, C. Yu, T. H. Ueng, S. Chen, B. J. Chen, K. J. Huang and L. Y. Chiang, *Toxicol. Pathol.*, 1998, **26**, 143.
86. T. Tsuchiya, I. Oguri, Y. N. Nakajima Yamakoshi and N. Miyata, *FEBS Lett.*, 1996, **393**, 139.
87. J. P. Kamat, T. P. A. Devasagayam and K. I. Privadarsini, *Chem.-Biol. Interact.*, 1998, **114**, 145.
88. H. Wang, J. Wang, X. Deng, H. Sun, Z. Shi, Z. Gu, Y. Liu and Y. Zhaoc, *J. Nanosci. Nanotechnol.*, 2004, **4**, 1019.

89. D. Pantarotto, J. P. Briand, M. Prato and A. Bianco, *Chem. Commun.*, 2004, 16–17.
90. P. Cherukuri, S. M. Bachilo, S. H. Litovsky and R. B. Weisman, *J. Am. Chem. Soc.*, 2004, **126**, 15638.
91. N. A. Monteiro-Riviere, R. J. Nemanich, A. O. Inman, Y. Y. Wang and J. E. Riviere, *Toxicol. Lett.*, 2005, **155**, 377.
92. D. B. Warheit, B. R. Laurence, K. L. Reed, D. H. Roach, G. A. M. Reynolds and T. R. Webb, *Toxicol. Sci.*, 2004, 77, 117.
93. A. A. Shvedova, V. Castranova, E. Kisin, D. Schwegler-Berry, A. R. Murray, V. Z. Gandelsman, A. Maynard and P. Baron, *J. Toxicol. Environ. Health, Part A*, 2003, **66**, 1909.
94. P. Leong-Morgenthaler, J. Horlbeck, F. Suter and F. Zulli, *In Vitro Toxicol.*, 1997, **10**, 129.
95. A. M. Smith, X. Gao and S. Nie, *Photochem. Photobiol.*, 2004, **80**, 377.
96. G. Xiaohu, C. W. C. Warren and N. Shuming, *J. Biomed. Opt.*, 2002, 7, 532.
97. A. M. Derfus, W. C. W. Chan and S. N. Bhatia, *Nanoletters*, 2004, **4**, 11.
98. M. Green and E. Howman, *Chem. Commun.*, 2005, **1**, 121.
99. A. Cahouet, B. Denizot, F. Hindre, C. Passirani, B. Heurtault, M. Moreau, J. Le Jeune and J. Benoit, *Int. J. Pharm.*, 2002, **242**, 367.
100. J. Zhou, F. Q. Zeng, C. Li, Q. S. Tong, X. Gao, S. S. Xie and L. Z. Yu, *Chin. Med. J.*, 2005, **118**, 50.
101. B. Wang, L. Zhang, S. C. Bae and S. Granick, *Proc. Natl. Acad. Sci. U. S. A.*, 2008, **105**, 18171.
102. J. McClements and D. J. McClements, *Crit. Rev. Food Sci. Nutr.*, 2016, **56**, 1334.
103. G. Singh, C. Stephan, P. Westerhoff, D. Carlander and T. V. Duncan, *Compr. Rev. Food Sci. Food Saf.*, 2014, **13**, 693.
104. P. N. Ezhilarasi, P. Karthik, N. Chhanwal and C. Anandharamakrishnan, *Food Bioprocess Technol.*, 2013, **6**, 628.
105. V. Stone and K. Donaldson, *Nat. Nanotechnol.*, 2006, **1**, 23.
106. T. Xia, M. Kovochich, J. Brant, M. Hotze, J. Sempf, T. Oberley, C. Sioutas, J. I. Yeh, M. R. Wiesner and A. E. Nel, *Nano Lett.*, 2006, **6**, 1794.
107. A. Nel, T. Xia, H. Meng, X. Wang, S. Lin, Z. Ji and H. Zhang, *Acc. Chem. Res.*, 2012, **46**, 607.
108. Y. Roiter, M. Ornatska, A. R. Rammohan, J. Balakrishnan, D. R. Heine and S. Minko, *Nano Lett.*, 2008, **8**, 941.
109. D. Fourches, D. Pu, C. Tassa, R. Weissleder, S. Y. Shaw, R. J. Mumper and A. Tropsha, *ACS Nano*, 2010, **4**, 5703.
110. T. Puzyn, B. Rasulev, A. Gajewicz, X. Hu, T. P. Dasari, A. Michalkova, H.-M. Hwang, A. Toropov, D. Leszczynska and J. Leszczynski, *Nat. Nanotechnol.*, 2011, **6**, 175.
111. B. J. Marquis, S. A. Love, K. L. Braun and C. L. Haynes, *Analyst*, 2009, **134**, 425.
112. W. H. Suh, Y.-H. Suh and G. D. Stucky, *Nano Today*, 2009, **4**, 27.
113. A. D. Maynard and E. D. Kuempel, *J. Nanopart. Res.*, 2005, **7**, 587.

114. C. M. Sayes, K. L. Reed, S. Subramoney, L. Abrams and D. B. Warheit, *J. Nanopart. Res.*, 2009, **11**, 421.
115. B. A. Magnuson, Nanoscale materials in foods: existing and potential sources. Intentional and Unintentional Contaminants in Food and Feed, *ACS Symp. Ser.*, 2009, **1020**, 47.
116. D. M. Kahan, D. Braman, P. Slovic, J. Gastil and G. Cohen, *Nat. Nanotechnol.*, 2008, **4**, 87.
117. D. M. Bowman and G. A. Hodge, *Futures*, 2006, **38**, 1060.
118. F. Cubadda, F. Aureli, A. Raggi, M. Cristina, B. Toscan and A. Mantovani, *ISTISAN*, 2016, **13**, 48.
119. N. O. Brien and E. Cummins, *J. Environ. Sci. Health, Part A: Toxic/Hazard. Subst. Environ. Eng.*, 2010, **45**, 992.

Subject Index